THE INTRODUCTION OF QUEEN BEES

Rege incolumi, mens omnibus una est.

VERGIL

Geor. IV, 212

This stately, beautiful, most noble and glorious Insect, in so many things expressing Royal Majesty, has been for a great number of years my grand favourite.

THORLEY

"The Female Monarchy", 1774

THE INTRODUCTION OF QUEEN BEES

By

L. E. SNELGROVE, M.A., M.Sc.

Fellow of the Royal Entomological Society
President of the Somerset Beekeepers' Association
A Vice-President of the British Beekeepers' Association
Expert and Honours Lecturer, British Beekeepers' Association
Member of the Apis Club

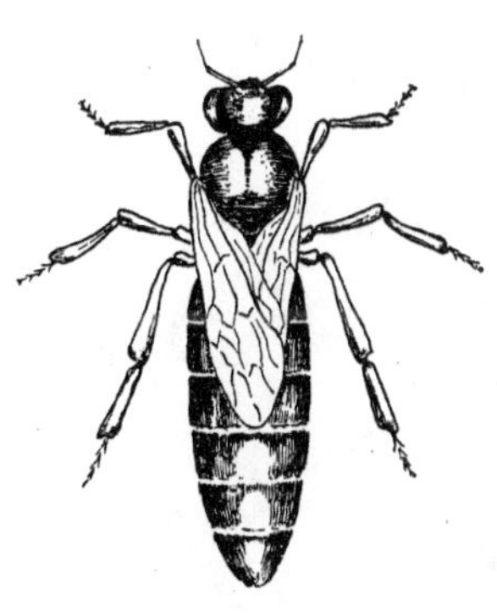

First Edition.   August 1940
Second Edition.   April 1943
Third Edition.   October 1948

PUBLISHED BY MISS I. SNELGROVE, BLEADON, SOMERSET.

MADE AND PRINTED IN GREAT BRITAIN BY PURNELL AND SONS, LTD.
PAULTON (SOMERSET) AND LONDON

INTRODUCTION

THE late G. M. Doolittle, famous American beekeeper
and pioneer of modern queen rearing, wrote at the
beginning of the century, "Perhaps there is no one
subject connected with beekeeping that has received
so much notice in our bee papers and elsewhere, as
has the introduction of queens." This is hardly surprising seeing that it is the beekeeper's method of
improving his stock. What is surprising, however, is
that notwithstanding its importance and the immense
amount of matter available—often contradictory—no
one has ever before attempted to sift it and write a book
on the subject. This Mr. Snelgrove has now undertaken
to do, and no one could be better fitted for the task.

The author is well known as a highly skilled practical
beekeeper, and an original worker with bees, possessing
a sound scientific education. He has already given us
many ingenious ideas and inventions, such as the
Snelgrove Board and System of Swarm Control, and
some original methods of Queen Introduction. He has
taken pains to investigate the early history of the
subject dealt with in this book, and to give careful
descriptions not only of his own work but also of all
the best methods of Queen Introduction in use. The
reader may be sure that the methods described have
been carefully tested in a scientific manner under
practical conditions, and that the judgments and
conclusions have been well weighed.

The practical beekeeper will be grateful to Mr.
Snelgrove for this valuable contribution to a most
important subject, and it is to be hoped that many
amateurs who have hitherto failed to introduce queens

successfully, or who have never attempted it, in the belief that it demands an amount of skill and experience they do not possess, will now be encouraged by this book to make the attempt. In it they will find valuable advice and a variety of methods suited to all circumstances.

L. ILLINGWORTH

Foxton,
        Cambs.

March 26th, 1940

PREFACE

My original intention was to write an account of certain new methods of Queen Introduction which I have devised in the course of experiments carried out during recent years. These methods are already widely used as a result of lectures which I have given. A description of them however could hardly be separated from a consideration of other methods or from certain matters relating to the general management of bees.

It is noteworthy that apart from two very small brochures, one by Raynor in 1884 and the other by Keen in 1938, there is no book in the English language, nor, so far as I have been able to ascertain, in any foreign language, which is specifically devoted to the important subject of Queen Introduction. A few of the largest works on Beekeeping, notably Cheshire's "Bees and Beekeeping", Root's "ABC and XYZ of Beekeeping", and Herrod-Hempsall's "Beekeeping, New and Old," contain sections in which the more usual methods are admirably described. A more comprehensive treatment of the subject appears in some sections of Perret-Maisonneuve's "L'Apiculture Intensive et l'Elevage des Reines". Most other writers dismiss the subject in a few paragraphs.

The improvement of stock by selection is of prime importance to all breeders of domesticated animals. In beekeeping this is effected by the renewal of queens, and this postulates a knowledge of safe means of queen introduction. Its importance is recognised by all beekeepers and need not here be emphasised. It is safe to say however that large numbers of beekeepers look upon the introduction of queens as so difficult, and fraught with so many risks, that they refrain from it altogether or are dependent on experts to undertake it for them. Failures in introduction, involving consequent loss of bees and honey, are far more numerous than are generally admitted, and are mostly due to imperfect appreciation of the conditions essential to success or to the inadequacy of the directions given in some textbooks. Sufficient directions to cover all contingencies cannot be given in a short paragraph, nor on a label attached to a queen-cage.

The reader will gather from a perusal of this book that whilst admitting the reliability of the older and more tedious methods of queen introduction I am an advocate of what are known as direct methods. In these days of rapid progress time is an important consideration. My own methods, effected in a few minutes or an hour at the most, are at least as safe as, and involve much less labour than the time-honoured caging methods which need the attention of the beekeeper for some days.

In addition to giving a fairly full account of my own work I have commented upon and criticised—impartially, I hope—the work of others. By means of numerous references I have endeavoured to attribute ideas and inventions to their originators, some of whom have been almost forgotten.

At the end of the book I have given a summary of the more important methods of queen introduction with an estimate of the value of each in practice. This summary is for the convenience of the busy beekeeper who may wish to select and employ a particular method in a hurry. It should be used only after a perusal of the fuller descriptions in the text.

I venture to hope that the book will prove to be of practical value to the average beekeeper and of interest to the scientific student and teacher of beekeeping; that it will be regarded as filling a void in beekeeping literature, and accepted as a contribution to apicultural science.

I take this opportunity of expressing my indebtedness to the following:—

(1) Miss M. Bindley and Mr. L. Illingworth, B.A., for assistance in consulting German bee literature.

(2) Messrs. R. M. Hopkins and G. H. Logan for the preparation of the illustrations.

(3) Dr. J. Wallace, O.B.E., M.B., B.Sc., and Messrs. L. Illingworth and W. Withycombe who read the proofs and made helpful suggestions.

(4) Messrs. G. H. Logan and F. W. Moore who kindly made apparatus needed for some of my experiments.

(5) Mr. W. H. Hermon who has confirmed the results of some of my experiments, and Mr. R. C. Cook who has given me much practical assistance.

L. E. SNELGROVE

Bleadon
March, 1940

CONTENTS

LIST OF ILLUSTRATIONS

CHAPTER I

BRIEF HISTORY OF REQUEENING

THE art of introducing queens, as we know it, has not been practised for much more than a century. It is a remarkable and most interesting fact however, that even in Roman times there were beekeepers who knew that it was possible to change the queen of a stock, and that in this way the disposition of the bees could be changed. Both Varro, who wrote during the first century B.C., and Columella, who wrote during the first century A.D., refer to requeening as though it were an ordinary practice. The former[1] says:—

"imbelliciores secretas subiiciunt sub alterum regem,"—

"they (the beekeepers) put the less vigorous bees apart under a new king"; the latter[2] writes:—

"domino mortuo . . . ex iis alvis quae plures habent principes, dux unus eligitur; isque translatus ad eas quae sine imperio sunt rector constituitur,"—

"when the king is dead a single leader is chosen from those hives which have several princes; transferred to the kingless bees he is set up as their ruler."

We are not told how the Roman beekeepers managed to introduce the queens. It is probable that they were often unsuccessful without knowing it, and that the bees raised their own young queens after destroying the queen or queen-cell introduced to them. When

[1] Varro, De Re Rustica, III, 16, 35.
[2] Columella, IX, 11, 3.

Columella speaks of hives with several princes he is probably referring to queen-cells or newly emerged virgin queens found in the hives a few days after swarming. In certain conditions of queenlessness the direct introduction of a ripe queen-cell or a virgin queen would be successful.

Purchas[1] in his book entitled "A Theatre of Political Flying-Insects," published in 1657, speaks of queen introduction as an ordinary process:—

"I once had a good swarm likely to miscarry, the Queen bees wings being accidentally torn, so that going out of the Hive shee fell beside the stool, and could not get up again. And although once and again finding her before the Hive, I put her into it, yet afterwards shee went forth, and was lost. And almost all the Bees (in her loss) neglected their work, and began to pine away, until I furnished them with a new Commander."

Purchas does not say how he furnished the new commander but in a later paragraph he writes:—

"If therefore you perceive a hive, after it hath cast twice, to have some quantity of Bees, yet to work negligently, or not to increase in the spring, suspect them to want a queen, and supply them with one as soon as you can, if no other way, by driving a poor swarm into them, for which purpose always reserve some".

Later he adds:—

"Nay, being necessitated, their own leader miscarrying in swarming, or by some accident afterwards, I have preserved the stock by putting to them a

---

[1] Purchas, S., *A Theatre of Political Flying-Insects*, pp. 30–32.

Queen Bee taken from another. And once, because I would bee sure, I pared off a little of one wing, and some months after, for experience sake, took the Hive, where I found that Commander put in by mee, and no other Leader".

The knowledge of queen-introduction possessed by the Romans does not appear to have been greatly improved upon until the 18th century.

In 1740 the celebrated Réaumur published his great work on the History of Insects, and in the portion devoted to bees he described several of his experiments on the introduction of queens. He used a direct method —merely placing a queen amongst the bees of a queenless stock and watching her reception through glass. He was sometimes successful and seldom refers to failures. He marked his queens with coloured paint, and on several occasions watched their favourable reception even when he introduced them to hives which had not been dequeened. In one case he successfully introduced two marked queens to a hive already containing a reigning queen, but later found that only one remained.[1] He ran queenless bees on to a group of queen cells and observed that a young queen duly emerged and was accepted by the bees.[2]

Réaumur's most interesting experiment was as follows:—[3]

Having a spare queen at his disposal he placed her in a small wooden box. This box, provided with a round hole which could be closed by a cork, may be considered to have been the precursor of modern queen cages.

During the month of April he took a small hive, fitted with glass sides, and provided with a wooden

[1] Réaumur, K. A. F., *Mémoires pour servir a l'Histoire des Insectes,* Vol. V., p. 639.
[2] Ibid., p. 271.
[3] Ibid., p. 258.

base in which was a round hole which could be closed by a cork. He placed this small observation hive, with the hole in the base opened, over a populous stock. When about 400 bees had entered he removed it, closed the hole, and took the imprisoned bees into his study. They soon showed great agitation. Whilst they were in this state the holes of the queen's box and of the beehive were uncorked, placed together, and the queen allowed to run into the hive. Hardly had she entered when a dozen bees surrounded her, showing pleasure at her presence and removing from her the dust with which she had become covered whilst in her box. Her "court" increased rapidly and for two hours she remained on the floor of the hive, surrounded, and at times covered, by the bees. They were given some honey, built a small piece of comb, and were liberated after two days. The bees were contented, and flew and returned in the usual way, but on the same day Réaumur noticed the queen outside at the entrance and re-introduced her. He concluded that she had been discontented with the small number of bees in the hive. Two days later they decamped, and were found, with the queen, as a tiny swarm on a branch of a plum tree. Re-hived, they finally disappeared,— doubtless as a starvation swarm.

By this experiment Réaumur had discovered that a strange queen is readily accepted by bees which have been deprived of their queen, brood, and food (p. 67).

Towards the end of the 18th century François Huber carried out his classical experiments in the course of his study of the life history of bees. Born in Geneva in 1750 he lost his sight when a young man, but this did not deter him from making an intensive study of the behaviour and economy of his bees which he kept in observation hives and studied through the eyes of his wife and his servant François Bernens.

In one of his experiments[1] he removed the queen from a hive. The bees at first appeared to be unaware of their loss and continued to carry on their ordinary work, but soon an agitation commenced near the spot which the queen had occupied in the hive. This commotion increased until the bees had ceased their work and were rushing over the combs in all directions in frantic search for the missing queen. As soon as she was restored to them the turmoil ceased instantly. Huber then repeated the experiment, substituting a strange queen for the original mother queen. He found that if a strange queen were introduced to the queenless bees within twelve hours she was imprisoned within a cluster or "ball" of bees where she died from hunger or suffocation. If introduced after eighteen hours she was "balled" at first but subsequently liberated and accepted as the reigning queen. If however the introduction were delayed until twenty-four hours had passed the strange queen was well received and allowed to reign.

Although Huber had made this notable contribution (amongst many others) to bee knowledge, it was not until the middle of the 19th century, with the invention of horizontally-movable frames by Baron von Berlepsch in Germany, and of vertically movable ones by Langstroth in America, that his discoveries could be turned to much practical use. Hitherto beekeepers had relied mainly on natural swarming for the renewal of queens, but now it was easily possible to de-queen a hive, and to introduce a new queen after sufficient confinement in a cage exposed to the queenless bees. The practice of re-queening gradually became an essential feature of good beekeeping. Ingenious queen cages were invented for this purpose by many people, notably by Alley, Benton, Doolittle and Miller in America, and by Raynor and others in England.

[1] Huber, F., *Nouvelles Observations sur les Abeilles*, Vol. 1, p. 190.

The idea of queen cages was not new however for Huish, writing in 1817, says:—

"The German apiarians are so fully convinced that it is by the particular smell that Bees recognise a stranger, that they always adopt the following plan on giving a queen to a hive: they confine her in a little cage which is placed in the hive, and in which she is suffered to remain about three days. She is then set at liberty, and the Bees accept of her as their monarch."[1]

The introduction of Italian bees to other countries began soon after 1850 and as it was found extremely difficult to transport whole colonies the practice of sending queens through the post developed, and cages suitable for both travelling and introduction were evolved.

According to Pellett[2] queens were probably first sent through the post in 1863. The first travelling cages contained only one compartment for queen, bees, and food. The food was a small piece of sealed honey comb, previously licked dry by bees, fixed to the interior of the cage. The queens were sometimes killed in the course of a journey through becoming daubed with honey. This difficulty was overcome by Mr. F. Benton who invented a travelling cage with separate compartments for bees and food, the latter being in the form of stiff candy which did not stick to the bees or queen when they came into contact with it.

Modified forms of the Benton cage are still in general use, and the method of introduction of queens by caging, notwithstanding the time and labour involved, is still generally recommended in text books as the safest and most suitable for the ordinary beekeeper.

[1] Huish, R., *A Treatise on the Nature, Economy, and Practical Management of Bees*, p. 211.
[2] Pellett, F. C., *History of American Bee-keeping*, p. 94.

Many advanced beekeepers however, who have little time to spare, prefer more direct methods as saving time and labour both to themselves and the bees, and it is the purpose of the writer to describe not only caging methods but also certain direct methods, including his own, which are eminently satisfactory.

There appears to have been no recognised method of direct introduction until the latter part of the 19th century, but there is evidence that individual beekeepers occasionally attempted to find one. The following is an interesting account of an experiment which, like that of Réaumur more than a hundred years earlier, if followed up, might have led to what is now a valuable method of direct introduction (p. 67).

In 1864 the Rev. Dr. Cumming wrote a series of letters on Beekeeping to "The Times" newspaper under the signature of "A Bee-master". Numerous letters were received in reply and amongst them the following from the Hon. and Rev. Orlando Forrester[1] :—

"Last autumn I saved from a cruel and unnecessary death a very old stock of bees belonging to a neighbour, the parent one of all his colony, full of black comb; I gave him seven or eight shillings for it. I fed it, and it just lived through the winter. I gave it a box on the lateral plan, which it filled with a quarter of a hundredweight of honey—so rewarding my interference. At the time I took my box from this stock, a neighbour destroyed two of his hives, bees, etc. for the honey (I think he will not do so ungrateful a thing another year). Happening to go into his garden the morning after, he told me he had found the queen that morning dead. On showing her to me, we found she was alive, but of course none the more lively for the sulphuring of the previous night. I begged her, took her home, and put her under a

---

[1] *The Times* Bee-master, *Beekeeping*, p. 216.

finger-glass with a little honey. I then got down from my store-closet my box of honeycomb untouched, cut away a good bit of honey, leaving sixteen or seventeen pounds. I carried it back to its old spot, opened the communication, and soon had a good number of the little manufacturers in their old quarters. Towards night I closed the communication again, stopped them in, and carried the old hive away about fifty yards. This done, I opened the ventilator at the top of the box, and inserted the strange queen, and put the inverted finger-glass over to see the effect. They seemed in a moment to be in a strange commotion, apparently receiving her majesty with hurrahs, as the popular candidate at an election is received. The next morning they seemed quite satisfied, and I noticed pollen taken in, and a struggle once or twice with a drone which had remained— there may have been seven or eight of these gentlemen in my experimental colony. Of course, the carrying in pollen was soon discontinued, as there was no brood to feed; the material, weighing sixteen pounds, being all honey and wax. I also saw them active in defence of their homes against the wasps; and although some wasps contrived to enter, I have seen the bees bring out their corpses now and then since, showing that they appear right."

It will be observed that the writer first made the bees queenless and broodless, and therefore hopeless, and that he inserted the queen "towards night"—all essential conditions for her favourable reception.

It was not till 1881 that Simmins[1] first used the expression "Direct Introduction" and described two methods devised by himself, both of which are still used by experienced beekeepers.

From this time the development of commercial

[1] Simmins, S., *A Modern Bee Farm*, p. 282.

queen-rearing was accompanied by improvements in methods of queen introduction and by the invention of a great variety of queen cages—mostly of the Benton type. In 1909 Asprea announced his discovery of the value of queen excluder in the construction of an introducing cage. His principle, supplemented by that of Thomas Chantry, is a feature of the best introducing cages now in use.

In 1911 the writer devised the "Water Method" and in 1936 the "One Hour" Method. By these the amount of time, labour, and equipment necessary for the introduction of a queen are greatly reduced.

CHAPTER II

IMPORTANCE OF REQUEENING

THE prosperity of a stock of bees depends primarily on the youth and fecundity of its queen. The rate at which she lays fertilised eggs during the spring and early summer governs the subsequent strength of the hive population, and on this, in turn, depends the amount of surplus honey which may be stored in a particular season.

A queen usually becomes mated during the first two or three weeks of her life and may then live for four or five years during which time she lays eggs at a diminishing rate. At the time of mating she receives the semen of the drone which contains many millions of minute filamentous organisms called spermatozoa. These are stored in a small globular sac called the spermatheca whence they are liberated in small numbers to fertilise the eggs as the queen lays them[1]. Eggs so fertilised produce female bees—i.e. "worker" bees and queens. If however the queen withholds spermatozoa from any of her eggs—which she is able to do—these unfertilised eggs produce male or drone bees. As long as the store of spermatozoa is plentiful a queen can maintain a high rate of worker-brood production, but as it diminishes her egg-laying powers wane and she tends to lay an increasing proportion of drone-producing eggs. As drones do not gather honey or pollen, but are great consumers of bee-food, it is obviously to the advantage of the producer of honey that their numbers be kept as low as possible except in stocks where they are specially reared for mating.

[1] Betts, A. D., *Practical Bee Anatomy*, p. 43.

In nature the renewal of queens is effected in three ways—by natural swarming, by supersedure, and occasionally by the rearing of a successor if a queen dies of disease or is otherwise lost. Normally during early summer the queen leaves the hive with a swarm to found a new home. Her place is taken by one of a number of young queens which emerge from their cells about a week after the swarm has departed.

In the following year the original queen departs from her new home with a swarm, and again in the third year. By this time her fertility has decreased and in the fourth or fifth year, and often much earlier, her stock becomes weak in numbers because she lays comparatively few worker eggs. The bees realising that she is failing then make provision for the continued existence of the stock by raising a young queen to supersede her. This young queen is reared under ideal conditions, being raised from the egg, and receiving adequate nourishment during her larval stage. In due course she becomes mated and fertile. As she takes her place at the head of the stock the old queen either dies of starvation or weakness, or is killed by her daughter. It is not uncommon however for both queens to be found laying side by side for some days or weeks before the old queen dies.

Occasionally, when supersedure takes place in the late autumn, two queens may live side by side during the winter[1].

Queens raised under the swarming impulse, or for supersedure, are usually very good, but modern bee-keepers cannot rely on these natural means of requeening, first because they desire to prevent swarming, and secondly, because supersedure usually means at least one unprofitable season on account of the retention of the old queen until she begins to fail. Moreover, supersedure sometimes occurs at an inconvenient time of year.

[1] G. di Bene, *L'Apicoltore d'Italia*, 1938, p. 50.

Queens obtained by supersedure are usually considered by beekeepers to be better than others, but experience shows that some at least have no claim to excellence. Turner[1] urges that supersedure does not produce a good queen because she is reared from the egg of a failing queen, which he says, cannot be as good as an egg laid by the same queen in her prime.

This theory, based on empirical and limited observation, cannot be accepted without more convincing evidence than Turner was able to give. It is inconsistent with the observed superiority of many "supersedure" queens, and although it is considered frequently to apply to human beings it is contrary to the general experience of breeders of agricultural stock who, the writer is credibly informed, do not find deterioration in the offspring of pedigree animals as these advance in age.

As a rule a queen lays well during her first and second years but a diminution of fertility is usually observable in the third year. The length of time during which she is fully productive is determined by three main factors. It will vary:—

(1) directly as the amount of spermatozoa derived from the drone, and the rate of their use.
(2) inversely as the amount of stimulative feeding to which she may be subjected.
(3) inversely as the capacity of the brood-nest of the hive in which she is kept.

Amongst other conditions which have some bearing on the useful life of a queen are heredity, rearing, health, seasons, and management.

It is usual to assume that fecundity is a transmissible quality in queens, but on account of numerous variable circumstances, especially of feeding and of the uncertainty of male parentage, it is difficult to prove

[1] Turner, N. H., *Scottish Bee Journal*, October, 1938, p. 154.

this. It is certainly true that the daughters of a prolific queen are not always themselves prolific, and that a poor queen may be succeeded by a prolific daughter when the latter is reared under the conditions of natural swarming or supersedure. Beekeepers are probably on safe ground when they breed from their most prolific queens provided they have due regard for disposition and industry—qualities which can be more easily shown to be transmissible.

It is probable that the feeding of a queen during her larval stage has more bearing on her subsequent fertility than heredity. A female larva, which emerges from an egg at the beginning of the fourth day after the egg is laid, becomes a worker-bee or a queen according to the manner in which it is fed by the bees. If it is destined to become a worker it is fed on brood food, a glandular secretion of the nurse bees, for three days, and for a further three days on the same food mixed with honey and pollen. In other words it is partially "weaned" during the second three days.

If it is to become a queen it is fed on secreted brood food, without the addition of honey and pollen, in increasing quantities throughout the six days.

Snodgrass[1] however, quoting the studies of Nelson and Sturtevant says—"A larva given a large amount of food at hatching needs little extra food for two days. After the second day, not after the third, as formerly held, the worker larva is given undigested pollen grains and honey, and for the rest of its life it is fed mainly on these substances."

The copious supply of secreted food deposited in the cell for the queen larva is known as "royal jelly", and this and the food given to all larvae during their first three days are commonly regarded as one and the same thing. Experiments carried out by Von Rhein[2],

[1] Snodgrass, R. E., *Anatomy and Physiology of the Honey Bee*, p. 172.
[2] Von Rhein, W., *Uher die Entstehung des weihlichen Dimorphismus im Bienenstaate*, p. 664.

however, who fed larvae artificially in an incubator, using brood-food of the first larval stage, brood food of the second larval stage (containing honey and pollen), and royal jelly, on different groups of larvae, showed, amongst other things, that

(1) Brood-food of the first larval stage (without honey and pollen) was not sufficient to bring larvae to the pupating stage.

(2) Worker brood-food of the second larval stage fed liberally to larvae throughout their existence produced over-sized workers with extra-large ovaries and well developed spermathecae.

(3) Royal jelly alone given to larvae from their 2nd to 3rd day onwards produced only workers.

The fact that Von Rhein successfully reared larvae on royal jelly but failed to do so on worker brood-food showed that these are not identical.

Since the bees themselves are able to rear queens from larvae of 2 to 3 days the results of these experiments suggest that other factors, apart from the known difference in feeding, may be involved in the development of queens and workers from young larvae.

There is some difference of opinion as to the limits of age within which a female larva can be developed into a satisfactory queen. Langstroth[1] says that "When a colony is deprived of its queen the bees soon raise another, . . . they select, first, some of the oldest amongst those whose milky 'pap' has not yet been changed for coarser food. Such a selection is wise, for the older a larva is, the sooner the colony will recover a queen". As the milky pap was supposed to be changed for coarser food on the 4th larval day Langstroth evidently considered that good queens could be reared from larvae of two or three days of age.

---

[1] Langstroth, L. L., *The Hive and Honey Bee*, p. 259.

Dr. Miller[1] considered that bees could raise queens from larvae more than three days old.

Doolittle[2], Maisonneuve[3] and Herrod-Hempsall[4], give 36 hours as the limit of age of larvae from which good queens may be bred.

Meier, corroborated by Zander (cited by Maisonneuve),[3] observes that important changes in ovarian development occur in the queen larva at the beginning of the second day and Zander concludes that from that moment it is impossible to change the larva of a worker into a perfect queen.

Von Rhein[5] concluded from his experiments that weight is a better criterion than age in respect of the development of workers and queens from young larvae. When a queen larva exceeds 20 mg. in weight it cannot be turned back into a worker, and when a worker-larva exceeds 35 mg. in weight it cannot be developed into a queen.

Accepting the views of Meier and Zander we must conclude that queens should not be reared from larvae more than 24 hours old. Good queens can undoubtedly be reared from somewhat older larvae but who can say how much better these same queens would be if reared as such from the first larval day?

In cases of natural swarming and supersedure bees raise queens from eggs. It is when man interferes with nature that inferior queens may be bred from larvae of unsuitable age. Breeders, for example, may graft larvae of the second or third days into artificial queen-cells; a beekeeper may fail to realise that the "comb of young brood" from which he rears his queens may contain only unsuitable larvae, or, in conceivable cases,

[1] Miller, Dr. C. C., *Fifty Years Among the Bees*, p. 239.
[2] Doolittle, G. M., *Queen Rearing*, p. 30.
[3] Perret-Maisonneuve, A., *L'Apiculture Intensive et l'Elevage des Reines*, p. 72.
[4] Herrod-Hempsall, W., *Beekeeping, New and Old*, Vol. I, p. 640.
[5] Von Rhein, W., *Uber die Entstehung des weiblichen Dimorphismus im Bienenstaate*, p. 664.

eggs too young and larvae too old at the same time. In the latter case queenless bees might not wait for the hatching of the eggs, but would construct some queen cells over unsuitable larvae. The queens from these would be inferior, and, emerging early, would destroy any queen cells raised over the eggs.

Gilman[1] and others urge that queens should be reared only from eggs. This is sound advice, especially for the beekeeper who needs only a few queens, as the possibility of rearing from larvae of unsuitable age is eliminated. The commercial queen breeder who rears queens in artificial cell-cups does not find it practicable to transfer eggs from the cells in which they are laid. He therefore breeds from the youngest larvae and does not admit the contention that the resulting queens are inferior.

When a normal stock is deprived of its queen the bees almost invariably rear new queens from larvae of the first day. This has been proved many times in the writer's experience by the fact that the resulting young queens emerge from their cells early on the 12th day afterwards. Occasionally, however, the oldest emerges on the 11th day, or, very rarely, on the 10th day, showing that the bees in their haste have built one or more queen cells over larvae of the 2nd or 3rd day.

Queens reared in weak nuclei, in stocks deprived of nurse bees, or in stocks which are short of honey, pollen, or water, are usually inferior. Apart from the fact that such badly-reared queens may occasionally show visible imperfections such as under-size or defective heads or wings, their main characteristic is that they lay eggs slowly during the spring and early summer with the result that their stocks do not become strong sufficiently early to take full advantage of the honey harvest. Their useful lives are short and they are liable to early supersedure.

[1] Gilman, A., *Practical Bee Keeping*, Chapter IV.

The writer has referred to these points, not with a view to dealing with the subject of queen-rearing, but rather to account for the wide-spread existence of inferior queens and the consequent need for systematic requeening.

Hives of large breeding capacity, and seasons during which the weather is favourable to the prolonged production of brood, tend to exhaust the fertility of queens and hasten the time when they should be deposed.

Some stocks of bees are so vicious that they cannot easily be manipulated. Their bad tempers are eliminated by the introduction of fresh queens reared from stocks of milder disposition. Similarly stocks which are prone to excessive swarming or which are lacking in industry may be cured of these undesirable characteristics by requeening.

Requeening from healthy stock is an important factor in the treatment of certain diseases—particularly of Nosema Disease, and Addled Brood.[1]

In every apiary a proportion of stocks will be found in a backward state at the opening of the summer season. It is a waste of time and labour to preserve their queens and they should be requeened as early as possible.

To obtain a few early young queens for this purpose it is a good plan to force a very early swarm by stimulative feeding and restriction of the brood-nest of a good stock. The queen-cells should be distributed to nuclei, and as soon as the resulting young queens are laying they should be introduced to the backward stocks by one of the direct methods described in Chapters VII and VIII.

Stocks may become queenless from a variety of causes. The queen may die of disease or old age, or she may be accidentally killed during manipulation; she

____
[1] H. L. A. Tarr, *Brood Diseases in England*, p. 11.

C

may be balled and killed by the bees in the spring in consequence of untimely disturbance or robbing; or she may even be lost on the cleansing flight which queens sometimes take during spring or autumn.

When young queens are reared in late summer or early autumn, whether by the beekeeper or in the course of natural supersedure, they frequently fail to become mated on account of unfavourable weather or scarcity of drones. Such queens will become "drone-breeders" and their stocks, though strong in the autumn, will become progressively weaker and will ultimately die out unless they are early provided with fertile queens.

Young queens are often lost when taking their mating flights. Their stocks usually develop fertile workers and as the eggs of these produce only drones such stocks will perish within a few months unless they are requeened.

It is not worth while, of course, to provide a new queen for a very weak stock. This should be united to a stronger one, or destroyed if it is not healthy.

The modern beekeeper, therefore, if he desires the maximum of success, should not only renew his queens systematically, but he should also be prepared at all seasonable times to introduce new queens to stocks which may have become queenless or which for other reasons are unprosperous. He should be familiar with the underlying principles of queen introduction by the best and safest methods, and with the detailed steps and precautions to be taken in using them. . Large numbers of valuable queens are lost or ruined every year through the use of cumbersome or defective methods of introduction and frequently beekeepers themselves are unaware of their failures.

The great importance of selecting male as well as female stock needs but brief reference here. As drones are produced by parthenogenesis they can be improved:—

(*a*) directly by the selection of their queen mothers and

(*b*) indirectly and to a less degree by the selection of the drones of the second preceding generation— i.e. their grandfathers. The beekeeper who carefully selects his queens through several generations, and breeds drones in considerable numbers from superior stocks whilst repressing them in others, is doing all that is possible for the improvement of his male stock.

CHAPTER III

CONDITIONS AFFECTING QUEEN INTRODUCTION

THE introduction of a queen to strange bees is always a delicate operation and demands of the beekeeper a general appreciation not only of the conditions which are favourable but also of those which are inimical to success. He should realise that he is not dealing with an exact science in which the results of operations can be predicted with certainty, but with creatures whose reactions to variable conditions are uncertain. A chemist, for example, can predict that if he adds a solution of silver nitrate to a solution of sodium chloride a white precipitate of silver chloride will appear. The physicist can say that a heavy body falling in vacuo will attain a certain speed at the end of a given interval of time. The natural laws underlying these phenomena are immutable—there are no exceptions. The biologist, however, cannot formulate such rigid laws for his use but has to be content with approximate rules, especially when he is dealing with animal response to external stimuli. Thus people behave differently in the presence of a calamity; all dogs do not respond in the same way to the sound of a gun; some horses are more easily trained than others—and so on; and of the little creatures with which we are directly concerned it has been truly said—"bees do nothing invariably".

Like all other animals they are subject to inherent variations of disposition. They experience contentment in the midst of plenty, fear when disturbed, anger when provoked, anxiety when hungry or imprisoned, and distress if their queen is taken from them. They are sensitive to changes of temperature and weather, and are appropriately responsive to gentle or rough handling. In short they vary in mood or behaviour much in the same way as human beings, and if they are expected to accept an unusual or unnatural situation they must be approached at favourable times and often cajoled, deceived, or frightened, so that their natural reactions are temporarily repressed. The behaviour of creatures subject to emotional variation cannot therefore be formulated in a set of strict rules and so it is impossible to devise a method of queen introduction which will be infallible in all circumstances. By careful study of all the circumstances, however, and by modifications of our procedure to conform to them, we can make rules for queen introduction so nearly perfect as practically to eliminate risks of failure.

Every beekeeper knows that if the queen be removed from a stock of bees, and another queen be substituted for her at the same time without special precautions, the latter will almost invariably be killed. The bees realise that the new queen is an intruder and the queen knows that she is amongst strangers. Mutual and immediate hostility results in the queen being seized, encased in a "ball of bees", and ultimately stung or starved to death.

The reasons for this hostility and the means by which it may be prevented will now be discussed. They should be thoroughly appreciated by every beekeeper who practises queen introduction so that he may intelligently modify his procedure in varying circumstances.

1. QUEENLESSNESS.

It is hardly necessary to say that a queen should not be introduced to a stock already possessing a queen—yet this is sometimes inadvertently done.

If there be too long a period between dequeening

and requeening the bees will have raised queen-cells and the oldest of these may have matured before the beekeeper breaks the cells down. Even if he breaks them down at the proper time he may miss an inconspicuous queen-cell surrounded by worker brood. The resulting virgin queen will destroy a newly introduced one.

Sometimes a virgin queen fails to mate and the beekeeper, finding no brood in the hive three or four weeks later, and failing to detect the virgin queen, may erroneously conclude that the hive is queenless. It is usually fairly easy to find a virgin queen but it is seldom safe to say she is not there simply because one cannot find her. She is active, shy, and easily frightened, and is more likely to hide amongst the bees than a fertile queen.

If there is any doubt as to the presence of a virgin queen the combs may be shaken clear of bees and placed over an excluder. Only the bees will then ascend to the combs and the virgin queen, if there is one, will remain below the excluder.

In some seasons fertile queens cease to lay very early in the autumn, especially if the weather is continuously unfavourable. An inexperienced beekeeper, alarmed by the absence of brood when he makes his autumn examination, and failing to find a queen, decides that a stock is queenless and may wastefully introduce another queen.

If a beekeeper is in doubt as to the presence of a fertile queen, liberal feeding for a few days will induce her to recommence laying, and her presence will then be indicated by the appearance of her eggs. If on the other hand a comb of brood containing eggs from another stock be inserted into a hive suspected of queenlessness the absence of a queen will be confirmed three or four days later by the appearance of queen-cells on this comb.

2. PRESENCE OF QUEEN-CELLS.

When an ordinary stock is dequeened the bees begin to raise queen-cells after a few hours. If a newly introduced queen is liberated within three days the bees will destroy these cells. If advanced queen-cells are present the bees will occasionally retain them until one yields a virgin queen which will destroy the stranger. If however the latter begins to lay before the young queens are due to emerge, the queen-cells are usually broken down. To ensure success all queen-cells should be destroyed before a caged queen is liberated or an uncaged one introduced by a direct method.

3. PRESENCE OF FERTILE WORKERS.

When a colony has been queenless for a long time during the summer months some of the worker bees begin to lay eggs which produce drones. Without special precautions it is almost impossible to introduce to such a colony a fertile or virgin queen, or even a queen-cell; nor will it raise queen-cells from young worker brood given to it. It is important therefore that every beekeeper should know under what circumstances laying workers develop and also how to recognise their presence. They occur more frequently than is commonly supposed, and often, unperceived by beekeepers, cause the loss of valuable queens introduced to them (v. Chap. IV).

4. VIRGIN AND NON-LAYING FERTILE QUEENS.

Virgin queens are always much more difficult to introduce than fertile queens. Even the latter are not so well received if they have ceased laying for some time, as, for example, when they have been caged for many days or have made a long journey through the post. They then approximate in appearance, and probably in attitude, to virgin queens, and do not so eagerly solicit food from the bees as do queens "in full lay". It is probable, too, that their characteristic odours change when they cease to lay and that consequently they are less respected by the bees. Doolittle[1] states that "a queen taken from the mails runs around, provoking the bees to chase her, whilst a queen freshly taken from a hive does not."

This brings us to an important consideration. When caged queens have come from a distance through the post they should not be introduced by direct methods until they have been kept undisturbed for some hours in the dark in order that they may regain their composure. It is an advantage too if the hive to which a travelled queen is to be introduced is made queenless at least 6 hours, or better still 24 hours beforehand, each period terminating at dusk.

5. TIME OF INTRODUCTION.

Many authorities consider that the most favourable time for the reception of a queen is when the bees are gathering honey. This implies that other times are less favourable. Herrod-Hempsall[2] considers the early autumn to be the best time. Maisonneuve[3] says that if a queen is to be introduced at times when honey is not coming in, the bees should be copiously fèd in order artificially to produce the desired conditions. He thinks the months of June and August are equally favourable, as in those months calm usually reigns in the hives. He even advises beekeepers not to introduce queens after the 15th of September. He recommends times of moderate temperature, and absence of wind and storms, as conducive to the tranquillity of the bees, and prefers evening or night to the day. During an extensive experience the writer has been unable to find

---

[1] Doolittle, G. M., *Queen Rearing*, p. 75.
[2] Herrod-Hempsall, W., *Beekeeping, New and Old*, Vol. I, p. 677.
[3] Perret-Maisonneuve, A., *L'Apiculture Intensive et l'Elevage des Reines*, p. 264.

that any period between the beginning of May and the end of October is unfavourable to a carefully executed introduction, provided always that hot weather combined with scarcity of nectar has not induced robbing, and that there are no fertile workers.

During a normal season and in typical English districts robbing is prevalent in August, and fertile workers during July and August. Bearing this in mind the writer considers that the most favourable times for queen introduction are:—

(1) Spring and early summer when stocks are building up to strength.
(2) During a honey flow.
(3) During quiet weather in autumn when bees are still gathering nectar.

6. DIFFERENCE IN ODOUR.

It is commonly accepted that every stock of bees has its own characteristic odour, and that by it the bees distinguish between members of their own community and those of other stocks. The hive odour is strong and is perceptible to the human sense of smell, but the differences by which bees distinguish one another are not perceived by man. The hive odour is strongest within the hive and diminishes on exposure to air. This probably accounts for the fact that bees returning from the fields are often admitted into strange stocks. Once past the guards they quickly acquire the family odour and go unchallenged.

The characteristic odour of a stock of bees can be masked by the use of scents or other strong smelling substances so that bees are unable to distinguish one another, and this fact is utilised in uniting bees as well as in queen introduction. It can also be destroyed by immersion of the bees in water which enables a bee-keeper to introduce a queen directly to bees, or bees to a queen—a principle used by the writer for his "Water Method" (p. 130).

Young bees, newly emerged from their cells, probably have little or no distinctive odour and no ability to distinguish it in others. They are therefore not only received without challenge in any stock but readily accept any queen introduced to them. For this reason queens are sometimes introduced to stocks by caging them over emerging brood (p. 77 et seq.).

Whilst the odour of a queen must necessarily partake of that of the family it is probably different from that of her bees and variable in changing times and circumstances. A virgin queen newly emerged from her cell probably has no distinctive odour and is then easily introduced to a queenless colony. When she is a few days old she has developed an individual or colony scent and must be introduced with special precautions. A fertile and laying queen acquires a much more distinctive odour which is pleasing to her own bees and perhaps repellent to strangers. Herrod-Hempsall[1] says "the exhalation of a fertile queen is so intense as to be easily perceptible." Mr. W. Hamilton[2] advances the theory that the characteristic odour of a queen is diminished prior to swarming because she then receives less brood-food from the nurse-bees. He suggests that this change in odour may be a cause of swarming.

The odour of a stock of bees is reduced by exposure to air. For this reason Bro. Adam[3] and others recommend the removal of quilts for a time before introducing a queen by the nucleus method (p. 120). Similarly the scent of a queen diminishes when she is removed from her hive and kept alone for a time in a cage or box. This is one condition of success in certain direct methods of introduction (p. 125).

[1] Herrod-Hempsall, W., *Beekeeping, New and Old*, Vol. I, p. 678.
[2] *The Bee World*, January, 1932.
[3] Adam, Rev. Bro., Article on Queen Introduction, Somerset Beekeepers, Annual Report 1934, p. 19.

Incidentally it may be stated that it is advisable to render the fingers odourless by washing before handling queens or queen-cells. A queen-marker, cleaned with methylated spirit whilst in use, may make a queen unsuitable for immediate introduction.

The most commonly practised methods of introduction depend much on the reduction of queen and bees to the same odour. This is usually effected by caging the queen for some time, so that she cannot be harmed, in the presence of the bees to which she is to be introduced. Identity of odour however is not always necessary. Root[1] says that the queens of two hives can sometimes be exchanged by simply putting each in the other's place without disturbance to the bees, so that the latter have no suspicion of the change. When bees are thoroughly frightened and demoralised by much smoke, by dusting with flour, by drumming on the hive, by hopeless confinement when deprived of their brood, or by being shaken from their combs, a strange queen may be dropped amongst them and will be unmolested notwithstanding differences of odour.

The length of time during which a queen should be exposed to bees in order to acquire their odour and become friendly with them varies greatly according to circumstances and the method of introduction, and also, it may be said, in the opinions of different writers. Cheshire[2] says that 12 hours are sufficient in times of prosperity, whilst three days may be necessary late in the season. Doolittle[3] states that he often released his queens after 12 hours and found them all right, and that he rarely kept them caged for more than 24 hours. He adds, however, that he sometimes kept them caged for 10 days! Root[4] says a queen should be caged for 3 to

[1] Root, A.I. and E.R., *ABC and XYZ of Bee Culture*, p. 464.
[2] Cheshire, F. R., *Bees and Beekeeping*, Vol. II, p. 328 et seq.
[3] Doolittle, G. M., *Queen Rearing*, p. 79.
[4] Root, A.I. and E.R., *ABC and XYZ of Bee Culture*, p. 463, et seq.

6 days to acquire the hive odour. Herrod-Hempsall[1] gives not less than 24 hours for a fertile queen, and for a virgin queen 3 days. Maisonneuve[2] recommends 24 to 48 hours as the period favourable for acceptance. Cowan[3] advises from 10 to 12 hours, plus 1 to 2 days during which the bees liberate the queen. Some other writers advise liberation after 36 hours.

Such a variety of views is confusing to the ordinary beekeeper. The truth is that only a comparatively short time is necessary for a queen to acquire the hive odour and any subsequent hostility towards her is due to other causes. The writer[4] considers that when the queen is intimately associated with the bees one hour is sufficient for the acquisition of a common odour.

Gillet-Croix[5] makes the interesting observation that queens from the same mother have almost the same odour and that it is consequently easy to introduce one of them in substitution for another.

In all cases of queen introduction a preliminary period of queenlessness of one day, whilst not essential, is conducive to success (p. 21).

7. ROBBING.

Bees tend to rob one another's hives at all times except when nectar can be easily obtained from flowers. Providing the weather be sufficiently mild they will steal as readily in January as in August, as people who have used Frow's cure for Acarine disease in mild winter weather have often experienced to their cost.

The guards of a strong stock repel intruders, but those of a stock weakened in numbers or by disease may be unable to contend with them. Fighting at the entrance is later succeeded by a struggle within the

[1] Herrod-Hempsall, W., *Beekeeping, New and Old*, Vol. I, p. 692.
[2] Perret-Maisonneuve, A., *L'Apiculture Intensive et l'Elevage des Reines*, p. 288.
[3] Cowan, T. W., *The British Beekeeper's Guide Book*, p. 140.
[4] Chapter VIII, p. 135.
[5] Gillet-Croix, A., *Précis d'Apiculture et Sélection des Reines*, p. 48.

hive during which heat is developed and a state of general demoralisation induced. The robbers leave for their home (it is usually only one stock that robs another) at nightfall, only to return in greater numbers on the following day. This continues until the stock dies of starvation and exhaustion. Robbing begins with the admission of one marauding bee to a hive and often becomes chronic before the beekeeper notices it.

A hive that is being robbed is in the worst possible condition for the reception of a new queen. The bees are irritable and suspicious, and although those of the stock may have been suitably prepared to receive a new queen the robbers are not so prepared and are likely to attack her. The writer has proved that although a queen can be directly introduced with perfect safety to bees taken from one hive and confined for 5 minutes or more in a small box (p. 135), a bee introduced to the box from another hive, and therefore a stranger amongst the others, will pursue and attack the queen. The presence of only a single robber in a hive may therefore be sufficient to result in the failure of an otherwise perfect introduction.

Robbing is most prevalent after the honey season when foragers are numerous and active and little or no nectar is obtainable from the flowers. In the British Isles, August is usually the worst month for it.

When a stock is being heavily robbed it is seldom of any value and therefore should not be requeened. When slight robbing is suspected queen introduction should be so arranged that the liberation of a caged queen is effected during the night, or direct introduction carried out during the late evening when all robbers will have left the hive. To prevent the ingress of robbers on the ensuing days the hive, well ventilated and shaded, may be closed and temporarily placed in another position for a week. Imprisonment of the bees seems to be favourable rather than otherwise to success in introduction.

8. HUNGER.

Hungry bees are considered to be less likely to accept an alien queen than those which are well fed. If a stock is short of stores it should be well fed for at least a day or two before a queen is introduced to it. This is especially the case when caging methods of introduction are used. Hunger appears to make less difference when the best direct methods are employed but it is always prudent to feed beforehand if the bees are short of stores. The feeding should be discontinued for a few days immediately after introduction in order to avoid disturbance, notwithstanding that Maisonneuve[1] advises feeding the day before, on the day, and the day after introduction.

9. ATTITUDE OF QUEEN.

Whilst a queen is laying she solicits food frequently from the nurse bees in attendance on her. If kept away from the bees for more than 15 minutes she becomes sufficiently hungry to take honey from the cells if this is available. Although she can live for many hours without food it is not desirable that she be deprived of it for much more than an hour. When she is introduced to a strange hive by a direct method after a short period of hunger she shows no tendency to be hostile or to be afraid, but being concerned only with her appetite, her first impulse is to go to the nearest honey cell or to solicit food from the bees. This attitude goes far to dispel suspicion and to ensure her favourable reception. Simmins[2] utilised this principle in his well-known "fasting method" of introduction. He kept the queen in confinement, alone, and without food, for not less than half an hour before liberating her amongst the bees. As Hutchinson[3] well says "The condition

---

[1] Perret-Maisonneuve, A., *L'Apiculture Intensive et l'Elevage des Reines*, p. 264.
[2] Simmins, S., *A Modern Bee Farm*, p. 283.
[3] Hutchinson, W. Z., *Advanced Bee Culture*, p. 93.

and behaviour of the queen is very important. If she will only walk about upon the combs in a quiet and *queenly* manner and go on with her egg-laying, she is almost certain to be accepted if other conditions are favourable. Let her run and 'squeal', utter that sharp 'zeep, zeep, zeep', and the bees immediately start in pursuit".

10. DISTURBANCE.

Bees should not be made irritable by clumsy manipulation prior to the liberation of a new queen. Prolonged search for an old queen, the use of much smoke, the jarring of combs, the careless kicking of hive legs, or noisy replacement of lifts and roofs, all tend to annoy the bees and to put them on their guard against intruders. If they have been irritated, and especially if they are running excitedly about the combs, introduction should be delayed until they have settled down. As Langstroth[1] says "It is necessary, when the queen is released, that the bees be in good spirits, neither frightened nor angered, and there should be no robbers about, as they might take her for an intruder and ball her".

The writer has sometimes taken bees and caged queens over considerable distances by train or car for the purpose of demonstrating methods of direct introduction. Both bees and queens arrive in a nervous and excited state, and the latter, when introduced, show a tendency to run away from the bees and so to incite attack. When night-fall has come both bees and queens have regained their composure and introduction becomes easy.

11. DISPOSITION OF BEES.

Young bees are more favourable than older ones to the reception of a strange queen. Wedmore[2] considers

[1] Langstroth, L. L., *The Hive and Honey Bee*, p. 284.
[2] Wedmore, E. B., *A Manual of Beekeeping*, p. 22.

that a new queen is most readily accepted by bees up to 10 days of age, and then by bees of over 30 days provided the latter include some whose nursing functions have not been exercised. Bees of 11 to 29 days, being "well able to build queen-cells and in no desperate need of a queen" cause difficulties.

If a hive contains a preponderance of older bees, as in cases where they have been for some time queenless, the introduction of some worker brood, and sometimes prolonged caging (when caging is employed) are desirable. It is generally believed that vicious stocks are difficult to requeen. They are certainly more troublesome when manipulated but it is doubtful whether their inherent unpleasantness makes any difference in their attitude to a properly introduced queen. To facilitate the requeening of a vicious stock it should be removed to a new stand for a day, the flying bees returning to the old stand being housed in a temporary hive containing a comb of brood. The old queen can be easily found amongst the young and good-tempered bees and after the new queen has been safely introduced to them the two parts of the stock may be re-united (p. 124). The viciousness of the stock will gradually disappear as the progeny of the new queen takes the place of that of the former one.

Illingworth[1] gives the following interesting method of requeening a vicious stock:—

Remove the hive to a new position, about a dozen yards away. Set an empty hive on the old stand and provide it with empty combs, or partly combs and partly foundation. It will be all the better if some of the combs contain honey and pollen. No brood should be given.

The queen to be introduced is placed in a cage alone and the cage provided with candy which is

[1] Illingworth, L., *Bee World*, January, 1940, p. 2.

exposed so that the bees may liberate the queen by eating it.

Fix the cage between the central combs in the new hive, cover with quilts, and place a feeder over the feed-hole. The flying bees from the original stock will return to the old stand. Having no brood from which to rear a new queen they release and accept the caged one.

The original stock, containing only young bees not disposed to sting, can easily be dequeened and re-united with the new stock by the newspaper method (p. 124), or otherwise disposed of.

Some races of bees are disinclined to accept queens of other races. Certain strains of black bees, for example, are more impatient of the introduction of Italian queens than are Italian bees of black queens.

12. CONDITION OF BROOD-NEST.

It is a condition favourable to success in the introduction of a laying queen that brood in all stages, or at least eggs and young brood, be present. Her environment will then be consistent with her presence, she will at once be encouraged to lay, and the continuity of the life of the stock will not be broken. The case of a virgin queen is different. In natural circumstances, except in cases of supersedure, a virgin queen is not found in a stock containing eggs or young brood. When a virgin queen is to be introduced, therefore, only sealed brood should be present. Both fertile and virgin queens can, of course, be introduced to broodless bees, but the presence of brood of the appropriate age promotes the chance of success. A queen should be liberated as near to the brood as possible, as she will then be in the place where the bees would normally expect to find her. For this reason the central feed-hole is usually convenient for her introduction.

D

The compactness of the brood-nest is important. The brood combs should be restored to their original positions after dequeening. If combs of stores are inadvertently placed so as to divide the brood-nest into two separate parts it is possible for a queen to be accepted by the bees of one part whilst the bees of the other continue to raise queen-cells which may have been begun. A young queen from one of these will kill the newly-introduced queen even after the latter has begun to lay. Jay Smith[1] regards this as a common cause of failure in queen introduction.

The removal of all brood and eggs a short time before the introduction of a new queen creates a condition specially favourable to her reception. This expedient is useful in the case of a stock which persists in rejecting a new queen.

According to Maisonneuve[2] a queen is always refused in a hive which contains only drone brood. The writer however does not share this view, having found no difficulty in substituting a fertile queen for a drone producing virgin by the "One-Hour" method without the previous insertion of worker brood, which, of course would be a desirable precaution.

13. STRENGTH OF STOCK.

A strong stock is more inclined to attack an alien queen than a weak one and therefore a queen is more easily introduced to a nucleus than to a strong colony. Hutchinson[3] made use of this fact in the case of queens that had not laid for several days or that had arrived in a jaded condition. He introduced them first to nuclei and when they were "nicely laying" united the nuclei to strong queenless colonies. It is however unnecessary to take so much trouble, and the union of a nucleus with

---

[1] Smith, Jay., *Queen Rearing Simplified*, p. 91.
[2] Perret-Maisonneuve, A., *L'Apiculture Intensive et l'Elevage des Reines*, p. 267.
[3] Hutchinson, W. Z., *Advanced Bee Culture*, p. 95.

a strong stock in the manner described by Hutchinson is not always successful. A jaded queen should be at once introduced to bees as described on p. 135 and these should be fed with a little honey about 15 minutes later. She can later be put into a provisioned cage with her bees and subsequently introduced to a full stock by a reliable method.

It might appear strange that Maisonneuve[1] considers that weak stocks are less inclined to accept queens than strong ones. His reason is that the former do not cover all the combs and that an introduced queen may venture into unoccupied places. He recommends that a weak stock be reduced in size by a division board and that one should make sure in every case that the queen is not liberated on a comb without brood.

14. Premature Examination.

When a queen has been safely introduced she is not necessarily completely accepted by the bees. They remain suspicious of her, or she of them, until she has begun to lay well. After that she is finally accepted as the mother of the stock. If she has been imprisoned in a cage for a considerable time she may not lay for several days and if the hive be disturbed and examined during this period she may be attacked and killed. Exposure to light seems to be specially conducive to hostility in such a case. The effect of light on bees, living as they do in almost complete darkness, may be greater than is suspected. They probably recognise one another partly by sight. The writer shows in Chapter VIII that two queens may be safely kept together with bees in a matchbox for a considerable time provided that they are in darkness. If they are exposed to light for two or three minutes one queen will attack and sting the other. Various writers recommend that a hive

[1] Perret-Maisonneuve, A., *L'Apiculture Intensive et l'Elevage des Reines*, p. 269.

should not be disturbed for two, three, or four days after requeening.

Herrod-Hempsall[1] recommends an interval of at least six days.

Freudenstein,[2] speaking of a queen liberated from a cage by bees which have eaten through candy, says that she will join her people unfailingly if the beekeeper will restrain his curiosity for at least eight days and not stir up the bees by taking the combs apart.

A writer in *Imkerführer*, December 1937, says that in every case an inspection should be left until ten days after the requeening.

A safe rule, which the writer observes, is not to disturb a hive for a week after requeening.

15. BALLING.

If a caged queen be released prematurely in a stock she will be attacked and encased in a "ball" of hostile bees. During the encasement her wing membranes are often badly torn and although she may be liberated after a few hours and allowed to reign in the hive she is, in this respect, mutilated and liable to be superseded at any time. Usually, however, she is stung to death and cast out at the hive entrance on the following day. When a queen which is not very old shows ragged wings it may be inferred that she has been badly introduced.

Balling may be caused by unseasonable disturbance, especially in the early spring, or by premature examination after introduction. If a ball, which is usually about the size of a large walnut, be observed, it should be taken gently from the comb by the hand or a suitable vessel (e.g. a tumbler), and dropped into water. The queen should be rescued when the bees leave her, placed in a cage with a new escort from the same hive, (p. 135) and later re-introduced.

[1] Herrod-Hempsall, W., *Beekeeping, New and Old*, Vol. I, p. 692.
[2] Freudenstein, Dr. K., *Deutscher Imkerführer*, July, 1937, p. 100.

CHAPTER IV

FERTILE WORKERS IN RELATION TO QUEEN INTRODUCTION

REFERENCE has already been made (p. 39) to the presence of fertile workers as inimical to successful queen introduction. The average beekeeper knows little about the conditions under which they develop or the means of preventing them from making an appearance in a hive. He recognises the signs of their presence when they are laying but does not suspect that they may exist in great numbers as egg-bearing workers which have not reached the laying stage, and he may share the popular misconception that the number of laying workers in a stock is always very limited. Consideration of these matters in some detail is essential to all who would avoid some unexpected failures in queen introduction.

Although worker bees are females they are incapable of mating, and having no functional spermathecae,[1] cannot lay eggs which produce worker bees. They have rudimentary ovaries, however, in which eggs can be formed. Miss Betts states that whereas each of the two ovaries of a fertile queen comprise about 180 ovarian tubes, those of a worker bee have about 6 and those of a laying worker about 20. It was formerly believed that laying workers were those which had been reared in proximity to queen-cells and which during their larval stage had received an extra amount of the royal jelly or brood food which is so abundantly supplied to queen larvae. Fertile workers appear, however, in colonies which have not been allowed to raise queen-cells.

[1] Betts, A. D., *Practical Bee Anatomy*, p. 42.

Langstroth's[1] view was that their ovaries are enlarged because they were fed lavishly in the first stage of their larval development, and he adds that "the young bees thus raised, having no more larvae to nurse when the hive has suddenly become queenless, feed each other with their milky food, which excites their laying as it does for the queens."

It is generally accepted that laying workers may be expected to appear in a queenless and broodless hive, and that their appearance is always associated with conditions which promote the elaboration of brood food when there is little or no brood to be fed. Buttel-Reepen[2] expressed the view that laying workers have their origin in nurse bees which, when there are few or no larvae to feed, utilise as their own food the royal jelly they prepare, which leads to abnormal development of their ovaries and incites them to lay.

Koch,[3] writing of the impulse to swarm, which follows a diminution of brood-rearing, says:—

"There comes a shortage of unsealed brood so that the brood-food producers are no longer relieved of the richly-produced brood-food. The blood of the nurse bees is enriched by this food and so the ovaries become developed. . . . Dr. Leuenberger in Switzerland examined microscopically the ovaries of bees from drone-breeding stocks and proved conclusively that 70% of the bees, of all ages, showed ovaries with developed eggs. Tuenin in Russia found that even in a stock with a normally functioning queen there can be workers with developed ovaries, but that they lay no eggs as long as there is a queen and brood. Bees of any age can become laying workers and generally

[1] Langstroth, L. L., *The Hive and Honey Bee*, p. 77.
[2] Leuenberger, F., *Les Abeilles. Anatomie et Physiologie*, p. 163.
[3] Koch, K. K., *Deutscher Imkertag*, 1932, pp. 5–6.

appear 28 days after the loss of the queen. Their egg-laying lasts about 5 days and those that have been under observation laid from 19 to 32 eggs each. Experiments showed that a stock with much unsealed brood and new-laid eggs did not develop laying workers. When the queen slowed down in egg-laying the number of open brood cells diminished and the number of laying workers increased."

Some interesting experiments carried out by Tuenin[1] in 1935 showed

(1) That colonies which swarmed, as well as the swarms themselves, developed from 20 to 70 per cent of potential ("anatomical") laying workers;

(2) That laying workers were not found in stocks which did not swarm;

from which it appeared, he says, that the absence of larvae in the colonies which gave swarms and also the small quantity of larvae in the first days of the life of swarms, cause a modification in the feeding of workers and thereby the evolution of eggs in their ovarian tubes.

Tuenin further found that the feeding of bees with larval food and honey hastens the process of development of laying workers in comparison with bees fed with pure honey only.

Gontarski[2] observed laying workers in queen-right colonies. He concluded that they sometimes lay in the brood nest but that their eggs are removed by the nurse bees. He contested Tuenin's theory that they develop at swarming time because of a superfluity of brood-food, holding that they result from an unbalanced ratio of production and consumption in the average

[1] Tuenin, T. A., "Concerning Laying Workers," *Bee World*, 1926, p. 90.
[2] Betts, A. D., *Bee World*, 1938, p. 105.

workers, and that not only are they not a result of swarming conditions (as Tuenin holds) but that they themselves start preparations for swarming.

He observed that laying workers drifted into other queenless stocks where they were found to be laying four days after these stocks were dequeened. He therefore advises the abolition of laying worker stocks whilst queen-rearing is in progress.

Leuenberger[1] examined the bees of a stock which had lost its queen towards the middle of April. At the beginning of June he found that only 30 of 100 bees examined were normal, and that the remaining 70 had developed ovaries mostly containing eggs. These included bees returning with pollen to the hive, guards at the entrance, bees taken from the edges of the combs, and nurse bees.

It is probable that every strong stock, which from any cause has undergone a limitation of its brood, has a proportion of potentially fertile workers during the warmer part of the year and that these do not lay whilst a fertile queen is present or whilst there is a brood from which a queen may be raised. Like queens, they may solicit food from other workers whilst they are laying. Cheshire[2] describes how one of them received court in a ring of attendants. When he lifted her from the comb some of the workers refused to leave her, gathering on his fingers, caressing her, and offering her food.

There can be little doubt that the presence of potentially fertile workers in a normal stock is unfavourable to the reception of a newly-introduced queen. They cannot be detected by the beekeeper and the effect of their presence on queen introduction appears not to have been considered by any writer. It may well account for otherwise inexplicable failures when

[1] Leuenberger, F., *Les Abeilles. Anatomie et Physiologie*, p. 160.
[2] Cheshire, F. R., *Bees and Beekeeping*, Vol. II, p. 355.

ordinary care has been taken to ensure successful introduction.

The presence of fertile workers which are actually laying is definitely inhibitive to queen introduction, probably because, as Hommel[1] says, the bees have illusions that they are not queenless.

In both cases the risks of failure are reduced by the presence of brood, especially if this, or some of it, is unsealed.

In some cases within the writer's experience, laying workers have appeared in stocks within a fortnight after the loss of their queens. Usually they appear after a longer interval. In some stocks they do not appear at all, although potentially fertile workers may be present. It is therefore manifestly undesirable to leave bees queenless and broodless for more than a few days. Stocks and nuclei containing virgin queens should be examined weekly, for their young queens may be lost on mating flights and such loss is soon followed by the development of laying workers. The occasional insertion of a little young brood into queen-mating stocks will not only prevent the appearance of laying workers in the event of the queens being lost, but will also serve to maintain the strength of their populations.

At first there may be only one or two laying workers in a stock and the first eggs of these may easily be missed by an experienced beekeeper. A queen then introduced in the ordinary way will almost certainly be killed. Later there may be large numbers of them with abundant evidence of their presence. They cannot be distinguished from other bees by external physical characteristics. The writer has on two occasions detected them in the act of ovipositing but has not been able to observe that any special attentions are paid to them by other bees.

Their presence in a queenless colony is indicated by:—

1 Hommel, R., *Apiculture*, p. 435.

(*a*) Their eggs. These are scattered irregularly in the cells and frequently many are deposited in the same cell. They are shorter than the eggs of a queen and are not usually perpendicular to the bases of the cells when first laid, as are those of the queen, but adhere at various angles to the cell walls.

(*b*) Their sealed brood. As the eggs are mostly deposited in worker cells the resulting drones have insufficient room for development and the cappings of the cells are raised above the level of the neighbouring empty cells. The drones which emerge from these cells are smaller than ordinary drones.

(*c*) Pseudo queen-cells. Although the bees treat laying workers to some extent as queens and raise drone brood from their eggs, they seem ultimately to realise that they are queenless and persist in building queen-cells over some of the drone larvae. These are fed as though they were queens but ultimately die in the sealed cells. Such false queen-cells are often distinguished by their unusual length and smooth surface.

It is impossible to separate laying workers from the others. Formerly it was advised to shake all the bees from their combs at a distance from their hive and allow them to fly home, the theory being that the fertile workers, not having been accustomed to fly, would not find their way back.[1] The writer has proved that this practice is futile and that the laying workers (or some of them) return with the rest.

Lancini[2] recommends the bees be dispersed 20 metres from the apiary, and their combs with stores and brood (if healthy) given to other stocks. If it is desired to re-establish the lost colony, an artificial swarm, formed from strong stocks and furnished with a fertile queen, may take its place.

[1] Cheshire, F. R., *Bees and Beekeeping*, Vol. II, p. 354.
[2] Lancini, V., *L'Apicoltore d'Italia*, 1938, p. 245.

If a stock infested with laying workers is weak it should be united to a "queen-right" stock. The trouble will then cease as any excess of larval food will be needed for worker brood. If, however, the stock is strong it is worth saving and can be requeened without much difficulty (p. 99).

A young queen is usually mated and begins to lay during the first fortnight of her life. In unfavourable weather, however, mating is delayed and the first brood may not appear for a month or even more. It is sometimes observed that some of this late-appearing brood, when sealed, comprises a proportion of raised cells which obviously contain drones. The queen may then be considered by the beekeeper to be unsatisfactory and is perhaps destroyed by him. She should be tested for a time, however, for if the drone brood disappears, as it sometimes does, it is attributable to fertile workers. If drone-breeding persists the queen is useless.

QUEEN CAGES AND THEIR USE

WHEN a queen is removed from her stock and confined in a cage she needs air, room for movement, reasonable warmth, attendants, and food. If she is fertile she receives food from the tongues of her attendants, which in turn, feed on the store provided for them in the cage. If she is unmated she feeds directly from the store in the same manner as the bees.

If a queen is to be confined for a very short time, as in the course of manipulation, it is not necessary to provision the cage with food. Her attendants will keep her alive and well even though they become hungry themselves. If confinement is to last for much more than an hour however, food should be provided. In some cases of direct introduction food is deliberately withheld so that the queen may be hungry when liberated (pp. 125 and 141).

Much ingenuity has been displayed in the invention of queen cages, which, although diverse in form and construction, may be classified in three groups according to their purpose and methods of use, viz:—

A. Cages suitable for housing queens and their attendants for a considerable time, i.e. for several days. These are convenient for the provision of food, for travelling through the post, and for introduction.

B. Cages for confining queens, with or without attendants, on combs containing brood and honey.

C. Cages to contain queens only during the period of introduction. Some of these are provided with mechanical devices for releasing the queens.

Typical examples of each group will now be described.

Readers desiring to know of interesting modifications of some of those included in group A are referred to Herrod-Hempsall's "Beekeeping, New and Old," I, XI.

Directions for stocking and provisioning cages are given in Chap. VI.

Group A

COMBINED TRAVELLING AND INTRODUCING CAGES.

1. Benton Cage.

The best known, and in its modified forms, the most widely used queen cage is that invented by Mr. Frank Benton of U.S.A. about 50 years ago. Its essential characteristics have not changed and it has served as the model for many others.

It consisted of a block of wood about $4'' \times 1'' \times 1''$ in which three wells, two circular (W_1 W_2 Pl. II, Fig. 1), of $1''$ diameter, and one elliptical (W_3) $\frac{3}{4}'' \times \frac{3}{8}$, were contrived almost, but not quite, through the thickness of the wood.

These wells were spaced evenly apart and the intervening walls and one end tunnelled by a passage (P) of $\frac{3}{8}''$ diameter. This passage enabled the bees to move from one well to another and at the appropriate time to emerge from the cage.

The ventilation of the cage was secured by means of series of small holes (Vh) pierced in parallel grooves (Gr) on both sides of the cage and reaching into well (W_1). The other wells were not ventilated. The grooves served to prevent the holes from being accidentally closed by contact with other articles in the post.

Wells W_1 W_2 formed the compartment for bees and queen. W_3 and the passage leading from it to the end of the cage was filled with candy (p. 107). The candy hole at the end of the cage was covered by a piece of cardboard.

The cage was completed by a covering of wire gauze (G) and this covered by a lid (L) of thin cardboard or wood, each being separately tacked to the cage.

The ventilation of W_1 enabled the bees to some extent to regulate their temperature. If too warm they would occupy W_1. If too cool they retreated into W_2.

In modern patterns of the Benton cage the three wells are circular, the one used for candy being large or small according to the length of time to be taken in a journey. In some models the wall between the wells W_1 and W_2 is partially cut away so as to provide a larger compartment for the bees (Pl. II, Fig. 2). The passage P is tunnelled through the cage from end to end. The candy end of this passage is covered with thin cardboard (e.g. a piece of post-card), and the other end, which serves for the insertion of queen and bees, is covered with wire gauze. Ventilation of the queen compartment is secured by a saw-cut, Sc, on each side, penetrating wells W_1 and W_2. The candy well (W_3) is covered by a piece of cardboard (ABCD), or better still, a sheet of celluloid. The queen compartment is covered with wire gauze and the whole with a cardboard or wooden lid as in the case of the Benton cage.

For use in the apiary or for travelling short distances a small cage of this type is quite satisfactory. It has the following dimensions:—

Length $3\frac{1}{2}''$, Breadth $1\frac{1}{8}''$, Depth $\frac{7}{8}''$. Diameter of wells (W_1 and W_2), $1''$. Diameter of well (W_3) $\frac{3}{4}''$. Depth of wells $\frac{3}{4}''$. Diameter of tunnelled passage $\frac{3}{8}''$·

The larger cages are desirable if the bees have to travel for several days. They may be purchased from appliance dealers at low prices.

2. SNELGROVE CAGE.

The cage used by the writer is simpler and in some respects better than the foregoing. Its main dimensions

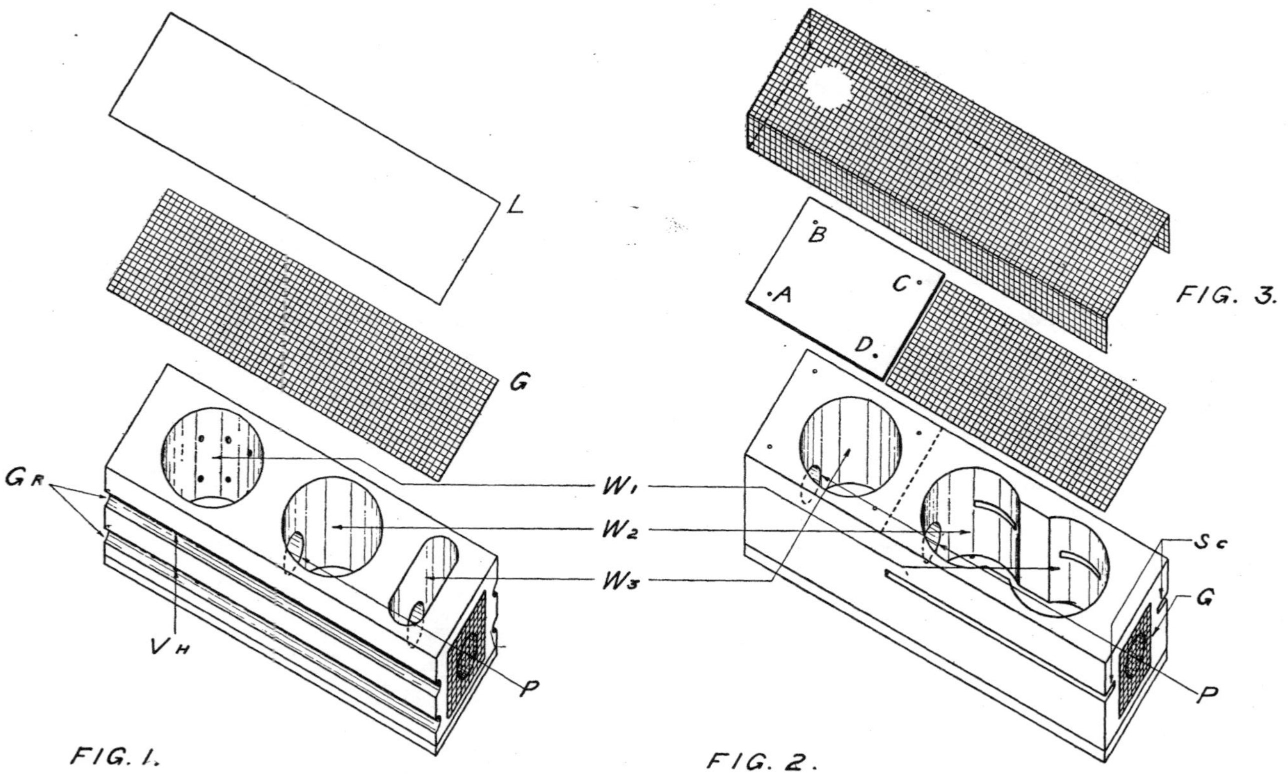

Fig. 1. Original Benton Cage.   Fig. 2. Modern Travelling and Introducing Cage.   Fig 3. Introducing Cover.

are $3\frac{1}{4}''$ x $2''$ x $\frac{7}{8}''$. There are only two wells—one for bees (W_1) and the other for candy (W_2). The former has a diameter of $1\frac{3}{4}''$ and the latter $\frac{3}{4}''$, each with a depth of $\frac{3}{4}''$. The candy-well is bored with a $\frac{3}{4}''$ brace-bit but the larger well is turned out by a lathe so as to give it a spherical surface (Pl. III, Fig. 1). The great advantage of this is that the queen can be lifted from the cage by the fingers if necessary, whereas in the case of ordinary cages it is impossible to insert the fingers into the wells and one has to wait until the queen comes out—which she is usually reluctant to do.

When the corners of the cage are rounded off it will drop into the feed-hole of a glass quilt, which is usually of $3\frac{1}{2}''$ diameter. This is an advantage as many modern beekeepers use glass quilts and find it impossible to insert an ordinary travelling cage either under the glass or into the feed-hole.

The cage is suitably ventilated for travelling. It is sufficiently large to house 20 or more bees with their queen and to contain enough candy for several days. The part containing the bees is covered with perforated zinc, and that containing the candy, together with the short tunnel, by a sheet of celluloid. This prevents evaporation from the candy and enables the beekeeper to see at a glance when the supply of food should be renewed.

The coverings are fixed to the cage by drawing pins which are more easily inserted and withdrawn than fine nails. By the removal of all the pins except one either covering can be swung round on the remaining pin so as to leave the cage partially or fully open.

3. GILLET-CROIX CAGE.

An interesting modification of the Benton cage is that devised by Gillet-Croix[1]. This contains two wells (W_1, W_2, Pl. III, Fig. 2), serving as a chamber for bees

[1] Gillet-Croix. A., *Précis d'Apiculture et Sélection des Reines*, p. 34.

E

and queen. These are connected with the outside by a tunnel (T) to contain candy and to serve as an exit for the queen when introduced. The principal feature of the cage is that it may be used for the "hatching" of a queen cell and subsequent introduction of the young queen. For this purpose the partition between the wells is cut away so as to leave a passage (P) between them which will accommodate a ripe queen cell. When the young queen emerges from the cell she cannot pass directly from one well to the other because the queen cell blocks the passage. Two passage-ways (Pw) are contrived in the side walls of the cage. They extend to the wells and also to the outside (O) where they are covered with wire-cloth. They enable the young queen to reach both wells and also to make the acquaintance of the hive-bees to which she is to be introduced. At the chosen time she is liberated by allowing the bees to eat through the candy in T.

The cage is provided with an additional well (W_3) to contain spare candy which is to be used to replenish the candy tunnel (T) after a journey, or before the queen is liberated. The bottom and sides of the cage are covered with wire-cloth and the top with a sliding lid of plain tin or zinc. It is suitable for every purpose for which queen-cages are used.

The central passage designed to accommodate a queen-cell is rather small and may need to be enlarged by cutting away the wood above it. When used for introduction by queen-cell the cage should be suspended between brood-combs with the tip of the queen-cell downwards. Metal supports are provided for this purpose.

4. THE MILLER CAGE.

The cage invented and used by Dr. Miller[1] differs from the foregoing in that it is sufficiently flat to be

[1] Miller, Dr. C. C., *Fifty Years Among the Bees*, p. 253.

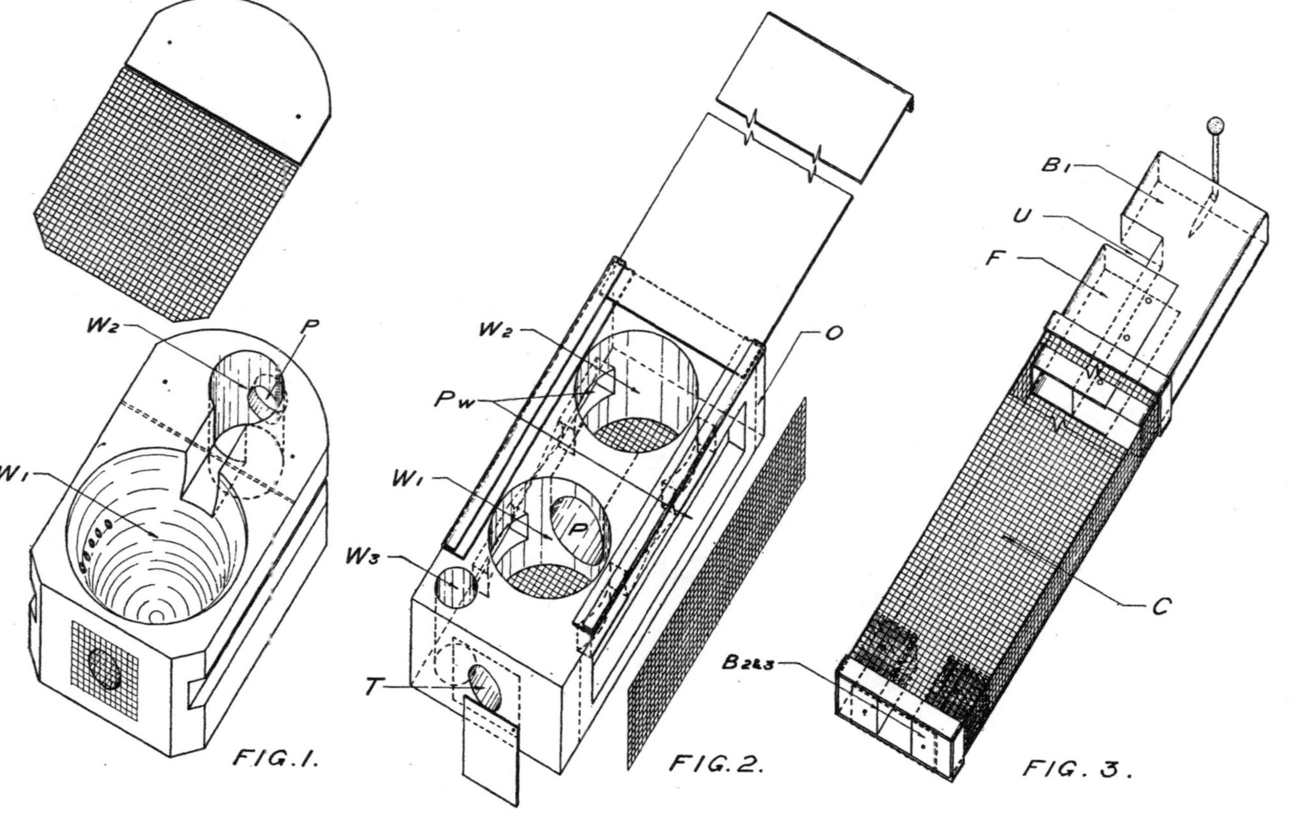

FIG.1. SNELGROVE CAGE.

FIG. 2. GILLET-CROIX CAGE.

FIG. 3. MILLER CAGE.

suspended between two combs without separating them widely, or to be pushed into the hive entrance. It is not suitable for transmission through the post.

It consists essentially of a wire-cloth cage (C) of rectangular section and of dimensions $4'' \times 1\frac{1}{4}'' \times \frac{3}{8}''$, made by folding the wire-cloth over a block of wood of slightly smaller size (Pl. III, Fig. 3). The open ends and one long edge are bound with tin for rigidity, and a block of wood (B_1) about $3\frac{3}{4}''$ long is made to fit and to slide fairly tightly into one end. Provision for candy is made in this wooden block by cutting away a portion, $1\frac{1}{2}'' \times \frac{3}{8}''$ from one side at one end, and partially enclosing the space so made by an arched sheet of tin, as shown at F. When this space is filled with candy the latter is available for the queen and bees inside the cage, and also—when the block is withdrawn sufficiently to expose the uncovered space (U)—to the bees of the hive. The position of the block therefore determines whether the hive-bees can reach the candy and liberate the queen.

In another pattern of the cage the sliding block is plain, and provision for candy is made at the other end of the cage which is left open in making. Two blocks (B_2, B_3), about $\frac{3}{8}''$ square and $1''$ long, are inserted and fixed in this end with a $\frac{3}{8}''$ space between them. This space is filled with candy and the end of the cage covered by thin card. The bees liberate the queen by tearing down the card and eating their way through the candy.

The use of the card covering for the candy makes the time of liberation of the queen uncertain. It is better to protect the candy with a metal covering and to remove this 48 hours after inserting the cage.

The cage may be suspended between two combs by means of attached wires, as recommended by Dr. Miller, or by a piece of flat metal fixed to the top of the wooden block and resting on the top bars of the frames.

This cage is simple in construction and convenient to use, but its chief disadvantages are that the hive must be opened and the bees disturbed when provision is made for the liberation of the queen, and that there is no secure retreat for her whilst exposed to hostile bees. It may be used for introduction at the hive entrance only when there is no likelihood of robbing.

For Latham's improvement of this cage see page 74.

INTRODUCTION BY TRAVELLING CAGE.

For the introduction of a queen by means of a cage of the Benton type a special movable cover of perforated zinc or stout wire gauze is sometimes used (Pl. II, Fig. 3). This is provided with a $\frac{1}{2}''$ hole (H) which, when the cover is in position, is over the candy. If the cage be moved away from the turned down end of the cover all three wells of the cage are covered by the zinc. The candy may therefore be protected from, or exposed to external bees by movement of the cage in its cover.

In use the cage with its cover is inverted and laid over a seam of bees between two combs, the cage being placed in the cover so that the candy is covered by the zinc. Alternatively, and preferably, the cage is laid on its side, the zinc face being vertical. Later, the cage is moved in the cover so that the hole in the latter is over the candy. The hive-bees then eat the candy away and liberate the queen.

To introduce a queen to a stock by means of a travelling cage of the Benton type it is necessary to expose the queen and her attendants, protected by the wire-gauze covering of the cage, for some time to the hive-bees. At first these are hostile but they gradually become friendly and begin to feed her through the gauze (p. 78). When this happens the candy may be exposed to them by the removal of the gauze (G) at the end of the cage. The time taken by the bees to liberate the queen depends on the amount and consistency of the candy remaining in the cage. If most of this has been eaten away by the caged bees more should be added so as to ensure, if possible, that the queen is liberated at night.

Goodacre[1] and Keen[2] describe a method of avoiding the uncertainty as to when a queen will be released by the consumption of candy. The exit hole of a cage is provided with a plug or cork which can easily be withdrawn by a fine wire attached to it. This wire extends to the outside of the hive. The cage, containing the queen only, is gripped firmly between the bars of the frames, and at the prescribed time the cork is pulled out and the queen liberated without disturbance of the bees.

The following methods of introduction by cages of the Benton type are reliable: -

(1) Dequeen the hive, and at the same time place the cage containing the new queen, with or without attendants, over the brood-nest, under the quilts. The cage should rest on its side so that the queen may retreat from the unwelcome attentions of hostile bees, and should be placed across the frames so that more than one seam of bees may become acquainted with her.

To prevent the bees from prematurely releasing the queen protect the piece of card at the candy end of the cage with a small piece of perforated zinc.

About 48 hours later—preferably in the evening—observe the attitude of the bees clustered on the face of the cage (p. 78) and if this is friendly, uncover the candy-hole at the end of the cage. If the candy has been consumed place a little more in the hole so that the queen will be liberated when the hive is undisturbed. Should the bees appear to be hostile re-examine the stock (Chapter III). If nothing wrong is found defer the exposure of the candy for another day. Replace the cage under the quilts and do not disturb the hive for a week.

---

[1] *The Australasian Beekeeper*, December, 1939, p. 190.
[2] Keen, C. H., *Introducing Queen Bees*, p. 10.

(2) Leave the hive queenless for 24 hours. Fill the candy well of the cage and the passage leading from it to the outside with stiff candy. Cover the candy-hole at the end of the cage with a piece of very thin card or thick rough brown paper and prick a few pinholes through it. Place the cage under the quilts as directed in (1). The bees will nibble their way through the card and eat the candy, and the queen will be released without further intervention on the part of the beekeeper.

(3) Carefully remove the wire cloth and the card covering the candy from the face of the cage and substitute for them the movable introducing cover described on page 70. Fix this in place with a drawing pin so that the candy and queen compartments are all covered by the perforated zinc. Place the cage, containing queen and attendants, zinc cover vertical, near a seam of bees, and leave for 24 hours (12 hours is usually recommended). At the end of this time remove the drawing pin, move the cage along to the turned-up end of the cover so as to bring the hole in the latter over the candy. Replace the cage over the seam of bees and do not disturb the hive for a week.

Various periods, e.g. 24 hours, 36 hours, 48 hours, and even 72 hours, have been recommended as necessary before the exposure of the candy for the liberation of a queen. The writer has always found 48 hours a safe and convenient period when the hive conditions are normal. Under abnormal conditions, e.g. when robbing is prevalent or when black bees are to receive a queen of one of the yellow races, it is advisable to keep the queen caged for three, four, or even five days and to destroy queen-cells before she is released (p. 39). The attitude of the bees indicates when it is safe to release her.

The queen may be caged in the hive without attendants. She then has to feed herself with candy and wait for proper food until the hive bees begin to feed her. The same objection holds if she is provided with an escort of newly-emerged bees. It is usual to introduce her with her original escort but for several reasons it is better to provide her with a new cage and candy, and attendants from the stock into which she is to be introduced (p. 135).

The "Asprea" principle.

In 1909 the Italian queen-breeder Dr. Asprea introduced a new and valuable principle in the introduction of queens by caging. He found that after a caged queen had been exposed to the bees of a queenless stock for from 12 to 24 hours the bees might be safely admitted to the cage through a piece of queen excluder fitted over one of the openings.

During the succeeding 12 hours the bees circulated freely through the cage and the queen could be safely liberated at the end of this time.

Asprea's principle however involved three visits to a hive. This is avoided in what is known as the Chantry principle which provides both for the automatic admission of the bees to the cage and the subsequent release of the queen.

The "Chantry" principle.

Root[1] describes the Chantry principle as applied to the Benton and other cages. A simple modified Benton cage, embodying this principle, is illustrated in Pl. IV, Fig. 1. The external bees gain admission to the cage by eating through two separate lots of candy, one a small amount contained in a short passage (P_1) near one end of the cage, and the other a larger amount filling a comparatively long passage (P_2) which also may terminate at the same end of the cage. To reach the smaller quantity of candy they must pass through a small piece of excluder zinc (Ex), affording one bee-way,

[1] Root, A. I. and E. R., *ABC and XYZ of Bee Culture*. Edition of 1929, p. 466.

which is tacked over the hole at the end of the cage. When they have eaten through the candy in the shorter passage, which they do in about 24 hours, they reach the queen compartment (Q) which extends from A to B and mingle peaceably with the queen and her attendants. They acquire the same odour, the queen is well fed, and the escort can escape into the hive. The queen however cannot yet escape from the cage, being prevented by the excluder. Meanwhile the hive bees will have gnawed away the piece of thin cardboard or paper (C) pierced with pin-holes, which covers the end of the longer passage containing the larger amount of candy, and will proceed to eat away this candy both from outside and inside the cage. This usually occupies them for two or three days, after which the queen is able to escape from the cage into the hive.

This cage, as used by the writer, is $3\frac{1}{2}''$ long and $2\frac{1}{4}''$ wide. The half which includes the queen compartment is covered with wire cloth and the other half with a sheet of celluloid.

An excellent all-metal cage, the "Ideal" (Pl. IV, Fig. 3) much used in U.S.A., is similar in principle. It has a sliding cover (C), side holes (H) for the insertion of the queen, and a suspensory attachment (A).

The principle is excellent because, on the assumption that all goes well, a queen may be introduced into a hive by one operation. It is used in the cages invented by Pieyre (p. 83), Adam (p. 84), Piana (p. 91), Mont-Jovet (p. 92) and Jay Smith (p. 92). Latham has applied it to the Miller cage by fixing a slot of queen excluder to the inside of the candy tunnel ($B_2 B_3$, Pl. III, Fig. 3) and substituting for the large movable block (B) a flat tunnel $2\frac{3}{4}''$ long made by tacking two sheets of tin to two strips of wood $3''$ long. The strips project $\frac{1}{4}''$ beyond the tin and their projecting ends are inserted into the cage when this is in use. The space between the strips is the larger candy tunnel.

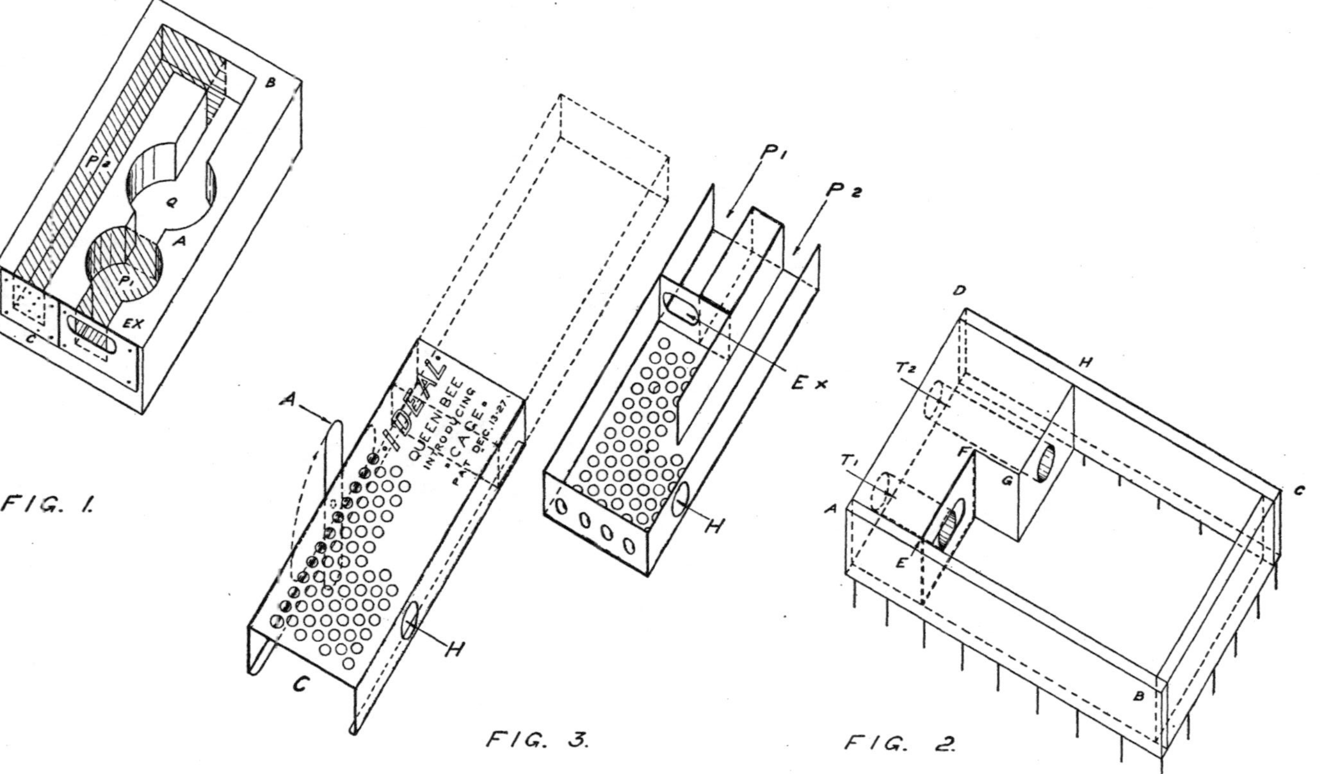

FIG. 1. CHANTRY BENTON CAGE.   FIG. 3. THE "IDEAL" CAGE.   FIG. 2. PIEYRE CAGE.

Both ends of the cage are filled with candy, the queen put in without attendants, and the cage suspended between two brood combs by means of a fine wire from a small stick laid across the top bars of the frames.

The cage is simple, easy to use, and effective.

A similar modification of the Miller cage has been devised and is used by M. I. Pritchard, of Medina, Ohio.[1]

Group B

CAGES APPLIED TO SEALED BROOD AND HONEY.

The distinguishing features of these cages are that the queens are kept in them only during the period of introduction and that they may have no attendants or provision of candy. The cages are affixed to tough comb so that the queens can feed themselves on honey until the hive bees are willing to feed them through the cage walls. They must not be pressed into new comb or the bees will eat through this, and reaching the queen prematurely, will kill her. The best known are what Root calls the "push into comb" cages. They assume many variations of form and are universally known and used. Some of them will now be described.

1. "Pipe-Cover" Cage.

The "Pipe-Cover" cage consists essentially of a cylinder of tin 2″ in diameter and 1″ in depth surmounted by a flat roof of wire-cloth (Pl. V, Fig. 1), or of a similar cylinder $\frac{1}{2}$″ in depth with a dome-shaped wire-cloth covering.

The cage containing the queen only is pressed to half its depth (i.e. to the depth of the cells) into a comb at a place where there are emerging young bees and also some cells containing honey. She is able to feed herself and soon has a retinue of young friendly bees.

[1] Root, A. I. and E. R., *ABC and XYZ of Bee Culture* (1940), p. 487.

If it is not possible to find emerging brood and honey together, the cage is placed over sealed brood and honey, and a few newly emerged bees are put into it with the queen.

The queen is liberated by hand after sufficient exposure to the hive bees. This involves disturbance unfavourable to reception. It is therefore necessary to observe the behaviour of the bees surrounding the cage before raising the latter from the comb. If they are quiet and attempting to feed the queen through the wire cloth, or, as Cheshire says, "paying court through the prison bars", the cage may be raised and the queen allowed to walk out on the comb. If the bees are then friendly they will back out of her way and touch her in an amiable manner with their antennae. It may then be assumed that she is accepted and the hive should be closed as quietly and with as little disturbance as possible.

If, on the other hand, the bees are unfriendly, a careful observer will notice that some of them cling closely to the cage and that one or more attempt to sting the queen through the cage wall. Should the queen be liberated in these circumstances, she will be pursued and held by one bee, to be quickly followed by others, until she is encased in a ball (p. 52). At the slightest sign of unfriendliness she must be caged for another day and the same caution again observed before she is released.

There is considerable diversity of opinion amongst writers as to the necessary period of detention of the queen by a "press-in" cage. Cowan[1] gives 24 hours; Maisonneuve[2] "24 hours, or better still 48 hours"; Cheshire[3] 12 hours in prosperous times, or late in the season, three days, and he quotes Root as saying without

---

[1] Cowan, T. W., *The British Beekeeper's Guide Book*, p. 136.
[2] Perret-Maisonneuve, A., *L'Apiculture Intensive et l'Élevage des Reines*, p. 288.
[3] Cheshire, F. R., *Bees and Beekeeping*, Vol. II., p. 333.

exaggeration that a week is sometimes necessary. Herrod-Hempsall gives 48 hours and this is doubtless a reliable period in normal cases, especially if liberation be effected late in the evening. Some authorities, including Benton[1] and Cowan[2] recommended sprinkling the bees with thin syrup when the queen is set free, but this practice is of doubtful value and at times might induce robbing.

It is to be remarked that all writers referring to the use of this type of cage emphasise the danger to the queen on liberation and insist on re-caging if the slightest hostility is shown. All danger is eliminated however if the cage is provided with half a dozen bees from the stock by the writer's method described on p. 135, the queen thereby being partially accepted before she is caged on the comb.

To place a queen and half a dozen young attendants in a "pipe-cover" or similar cage is not a very simple matter. The writer proceeds as follows:—

Cut a piece of postcard (C. Pl. V, Fig. 1) of the same width (2″) as the cage and about twice as long. Near one end cut a square or round hole (H) of about $\frac{1}{2}$″ diameter. Invert the cage and strap the card over it with a rubber band. The card can be easily moved along its length over the cage without dislodging the band. Move the card until the hole in it is over the cage. Take the cage in the left hand, placing the thumb over the hole in the card. The queen and bees can then easily be put into the cage one at a time. Slide the card again until the hole is clear of the cage. Place the cage, card downwards, on the face of the selected comb, gently withdraw the band and card, and with a screwing motion press the cage into the comb as far as the mid-rib of the latter.

---

[1] Cheshire, F. R., *Bees and Beekeeping*, Vol. II, p. 335.
[2] Cowan, T. W., *The British Beekeeper's Guide Book*, p. 136.

2. "AMERICAN" CAGE.

Root[1] describes a "push-in" cage, easily made of a rectangular piece of wire cloth by folding the sides and ends, which is used in the same manner as the "pipe-cover" cage, except that it is not pushed in to the mid-rib of the comb. The bees are able to nibble away the comb under its edges and release the queen within four or five days. Cowan[2] illustrates a similar cage, which he calls the "American" cage (Pl. V, Fig. 2). In this the lowest horizontal wires are removed, leaving the vertical wires projecting as spikes to a length of $\frac{1}{4}''$, which is the depth to which the cage should be pressed into the comb.

Maisonneuve[3] describes the same cage as recommended by Giraud-Pabou for the introduction of exhausted queens to emerging bees. Hutchinson's[4] method of liberating a queen from this cage is a good one. He bores a hole through the comb from its opposite side leaving the broken up comb in the hole. The bees clear this out and the queen is released without disturbance.

3. SMITH "PRESS-IN" CAGE.

The "Smith"[5] introducing cage devised by Jay Smith, a distinguished American beekeeper, provides for the retention of the queen on the comb until she begins to lay.

It consists of a rectangular wooden frame lined with tin, about $4'' \times 3''$, the top covered by wire cloth and the bottom open but provided with tin edges cut into teeth. These are to be pushed into tough comb containing, if possible, some honey, sealed brood, and

---

[1] *ABC and XYZ of Bee Culture.* Edition of 1929, p. 468.
[2] Cowan, T. W., *The British Beekeeper's Guide Book*, p. 138.
[3] Perret-Maisonneuve, A., *L'Apiculture Intensive et l'Elevage des Reines*, p. 305.
[4] Hutchinson, W. Z., *Advanced Bee Culture*, p. 95.
[5] Smith, Jay, *Queen Rearing Simplified*, p. 88.

PLATE V

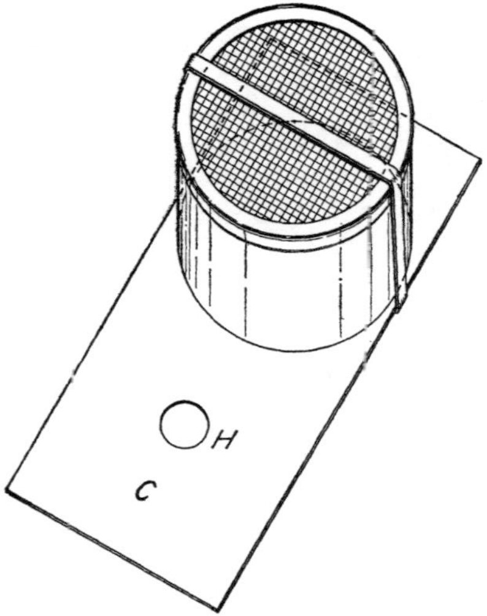

FIG. 1. PIPE COVER CAGE AND CARD.

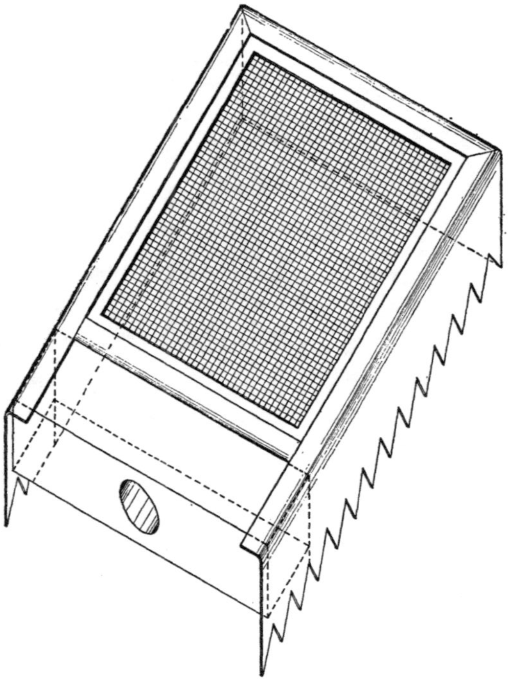

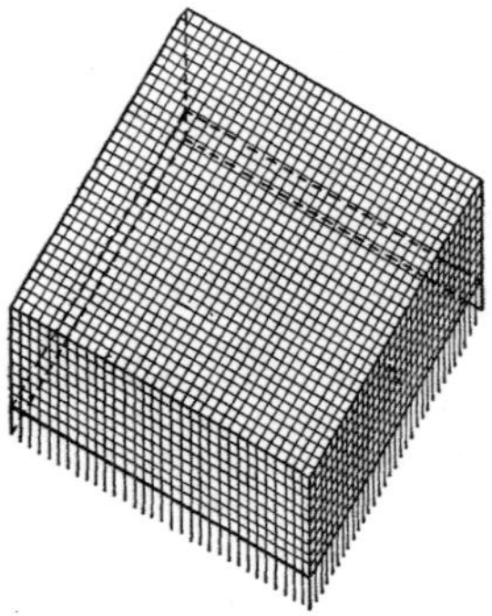

FIG. 2. "AMERICAN" CAGE.

FIG. 3. BOWEN CAGE.

empty cells. A corked opening in one end serves for the insertion of the queen after the cage is fixed to the comb. The queen, alone or with some attendants, is kept in the cage until she begins to lay—usually in 4 or 5 days—and is therefore amicably accepted when released.

The use of the Asprea principle (p. 73) is provided for in this cage, the reliability of which is thereby increased. A corked opening in one side, covered on the inside by excluder, serves to admit the hive-bees to the cage after two days. Two or three days later the cage is removed and the queen liberated.

The Bowen cage is similar to the Smith cage but is made entirely of metal and does not provide for the use of Asprea's principle. A wooden block (Pl. V, Fig. 3) pierced with a candy hole leading into the cage, serves for the insertion of the queen and enables the hive-bees to liberate her when the candy is exposed to them.

It must always be remembered that it is desirable that queen-cells be broken down before the release of a queen which has been caged in a hive for more than three days.

4. "PIEYRE" CAGE.

This cage, described by E. Alphandéry[1] in his charming book "Traité Complet d'Apiculture", and by whose kind permission the illustration (Pl. IV, Fig. 2) is reproduced, is an advance on the "push into comb" cages previously described in that it permits the use of the Chantry principle (p. 73) and therefore ensures the automatic liberation of the queen without the disturbance which is commonly found to be dangerous with other "push-in" cages.

The cage ABCD is made of thin wood, $\frac{1}{8}''$ thick, on three sides, and is $4\frac{3}{4}''$ long, $3\frac{1}{8}''$ wide and $\frac{3}{4}''$ deep. The

---

[1] Alphandéry, E., *Traité Complet d'Apiculture*, p. 230.

fourth side is a block of wood of the shape AEFGHD through which are bored two tunnels T_1, T_2 of $\frac{3}{8}''$ diameter, of lengths $\frac{3}{4}''$ and $1\frac{1}{2}''$ respectively, both reaching to the interior of the cage. The whole is covered with wire cloth and its lower edges provided with fine wire pins, $\frac{7}{8}''$ long, which are thrust into the comb when fixing the cage in place. The inner end of T_1 is covered with a small piece of queen excluder. Both tunnels are filled with stiff "Good" or similar candy, and, for additional safety, may be covered on the outside by very thin card pierced with pin holes.

In use the cage is placed over the queen alone on a part of the comb containing emerging brood. In a short time she begins to lay. Meanwhile the hive bees nibble away the card at T_1, consume the candy, and enter the cage one at a time through the excluder. They are friendly to the queen and pass freely in and out of the cage, but the queen cannot escape until the bees have been able to eat all the candy in the longer tunnel T_2. Pieyre states that nine times out of ten the queen will have begun to lay before she leaves the cage. He recommends that the hive be not disturbed for 6 or 8 days after the introduction.

It will be noted that whilst the principle of the Pieyre cage is good, the means of fixing it to the comb make it suitable to cover brood only. If placed over comb containing honey the bees would be able to burrow under the edges of the cage and liberate the queen prematurely.

5. ADAM CAGE.

A much better cage, similar in principle to that of Pieyre, is that devised and at one time used by the Rev. Bro. Adam.

This cage is provided with a separate queen compartment made in the block of wood carrying the longer candy tube. This compartment allows access to the candy from within, and also, by means of two passages, both to the exterior and the interior of the cage. These passages can be closed by metal slides.

The short candy tube, covered by a slot of excluder is bored through the opposite wall of the cage.

When the cage is to be used the small compartment is shut off from the large one by closure of a slide and the queen, alone or with attendants, is inserted into it.

The cage is now taken to the hive in which it is to be used and fixed to a suitable comb by means of a serrated tin rim.

The queen is then liberated from the small compartment by raising the slide and she and her attendants enter the larger compartment over the brood.

Both candy tunnels being comparatively short are guarded by thin pierced card on the outside of the cage.

The bees liberate the queen in the same way as described for the Pieyre cage.

For an illustration of the Adam cage see Herrod-Hempsall, p. 695.

Group C

CAGES TO BE SUSPENDED BETWEEN COMBS

1. CYLINDRICAL CAGE.

Perhaps the simplest of all cages and the easiest to make is that known as the "round" or cylindrical cage. A piece of perforated zinc or wire cloth $3\frac{1}{2}''$ square is bent into the form of a cylinder, the edges fastened together by fine wire or solder, and the open ends fitted with corks (Pl. VI, Fig. 1). The queen is placed alone in the cage and the latter suspended vertically between two combs so as to be in contact with sealed brood and honey. The combs are pressed slightly against the cage so as to support it. The queen feeds herself through the cage wall until the bees are willing to give her food. At the end of 48 hours (Bertrand[1] says 24 to 36 hours), the attitude of the bees is observed (p. 78), and if this is favourable the lower cork is removed and the queen allowed to walk out.

Halleux[2] recommends the substitution of a piece of comb honey for the lower cork and allowing the bees to eat through this to liberate the queen. (One concludes that candy would be preferable for this purpose.) If the bees appear to be unfriendly the queen is returned to the cage for another day.

There is some risk in the use of this cage apart from that caused by disturbance when the queen is released. It is to be pressed against sealed honey so as to bruise it. The bees are likely to eat a passage through this honey around the cage in a very short time with the result that the queen may be starved. Halleux[3] says that more than once he has found the queens dead at the bottom of these cages.

In the form of this cage used by Doolittle[4] short lengths of broom-handle were used instead of corks. The shorter, $1''$ long, was fixed. The other, $5''$ long, was withdrawable and had a candy hole $\frac{3}{8}''$ in diameter bored through its entire length. As the queen could feed herself from the candy, the cage was pressed between combs of brood. The hive bees liberated her by eating away the candy.

Alphandéry[5] suggests the cylindrical cage for the method of substitution. The old queen of the hive is first put into the cage for one or two days. The new one is then substituted for her and liberated, with the usual precautions, two days later.

[1] Bertrand, E., *La Conduite du Rucher*, p. 39.
[2] Halleux, D., *Le Livre de l'Apiculteur Belge*, p. 361.
[3] Ibid, p. 362.
[4] Doolittle, G. M., *Queen Rearing*, p. 84.
[5] Alphandéry E., *Le Livre de l'Abeille*, p. 38.

PLATE VI

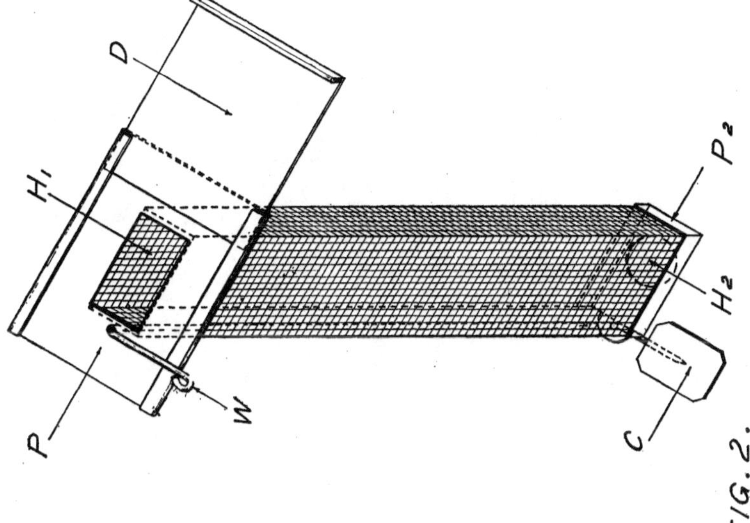

Fig. 2. Raynor Cage.

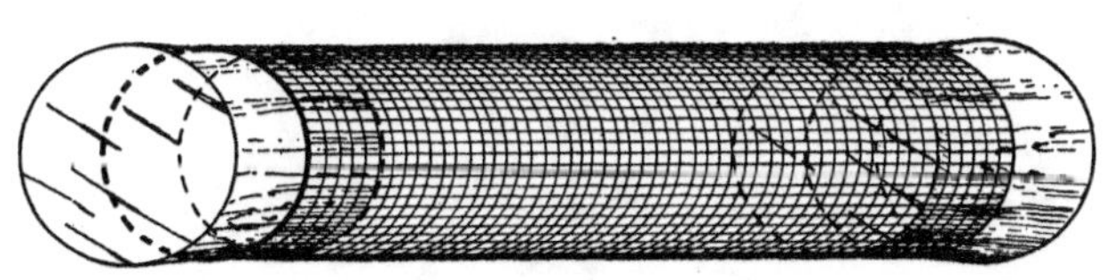

Fig. 1. Cylindrical Cage.

2. Raynor Cage.

A much better cage than the foregoing is that which, as Cheshire[1] says, "bears the honoured name of the Rev. G. Raynor"—a modification of a cage invented by "a Renfrewshire Beekeeper."[2] This is still frequently used and is illustrated in Pl. VI, Fig. 2. The body of the cage is a rectangular tube made of wire gauze of dimensions $4\frac{1}{2}'' \times 1'' \times \frac{1}{2}''$. It is suspended between two suitably spaced combs by means of a piece of tin plate (P) $2'' \times 1\frac{1}{2}''$, soldered to the tube. A hole (H_1) $\frac{7}{8}'' \times \frac{1}{2}''$ is cut in this plate to serve as an entrance for the queen to be introduced, and this hole can be covered by a tin door (D) sliding into the folded edges of the plate. The bottom of the tube is completed by another tin plate P_2 through which a round hole H_2 of $\frac{1}{2}''$ diameter is cut to serve as an exit for the queen. This exit can be closed or opened by a tin cover (C) fixed to a stout wire (W) exterior to and extending the whole length of the cage. The wire passes through the upper plate, which keeps it in position, and at each end is bent at right angles so that a rotary movement of it at the top opens or closes the exit hole at the bottom.

In Raynor's original cage the two doors were hinged and moved vertically, the lower being moved upwards or downwards by raising or depressing the wire.

Raynor used this cage for the method of substitution. He first caged the old queen of the hive and suspended the cage between two combs where the bees were numerous. After 12 hours he opened the top door and with a little smoke caused the queen to walk out. He then put the new queen into the cage and suspended it again between the combs for 24 hours, after which, provided there were no excitement at the hive entrance, he liberated her by depressing the wire. Presumably the

---

[1] Cheshire, F. R., *Bees and Beekeeping*, Vol. II, p. 329.
[2] Hunter, J., *A Manual of Beekeeping*, p. 128.

cage was pressed against honey so that the queen could feed herself until fed by the bees. Two important advantages of this cage and method are that the new queen, to some extent at least, acquires the odour of the old one, and that she can be released with a minimum of disturbance to the bees. Cheshire[1] advises that liberation be effected at night and suggests that the bottom of the cage be plugged with candy and the bees allowed to release the queen by eating it away.

Keen[2] recommends caging the old queen, without escort, in her own hive for 24 hours or more. The new queen is then to be substituted for her in the cage and the candy exposed so that the bees may liberate her. The risk involved in this procedure is obvious—the queen might be liberated before the bees had become friendly with her. It would be safer to expose the candy on the following day at such a time as would ensure the queen's release at night.

Sellner[3] advises preliminary caging of the old queen for four hours. The exchange is then effected and 24 hours later the attitude of the hive bees is observed. If this is friendly they are allowed to liberate her by eating through candy. He considers that an unfriendly attitude of the bees indicates that everything is not in order in the brood nest. (Chaps. III and IV.)

Pellett[4] refers to substitution as practised by some beekeepers when they expect queens to arrive by post. When they order the new queens they cage the old ones in their hives and leave them there until the newcomers arrive. The old queens are then destroyed, the new ones substituted for them in the cages, and these replaced in their respective hives. The cages having acquired the hive odours, the bees, being

---

[1] Cheshire, F. R., *Bees and Beekeeping*, Vol. II, p. 329.
[2] Keen, G. H., *Introducing Queen Bees*, p. 9.
[3] Sellner, Karl, *Natürliche Königinnenzucht*.
  Bienen Vater, Juni, 1937, p. 96.
[4] Pellett, F. C., *Practical Queen Rearing*, p. 96.

accustomed to the presence of queens in cages, have little suspicion of the new arrivals.

The method of substitution has the advantage of making the new queen more acceptable to the bees but it is rather troublesome and not to be preferred to introduction by cages embodying the Chantry principle.

3. THE "GAPIR" CAGE.

The "Gapir" cage, invented by the Italian queen-breeder Piana and used by him for many years, embodies the Chantry principle. The cage (Pl. VII, Fig. 1), $3\frac{1}{4}''$ long and $3''$ wide, is constructed of two wooden sides (S_1 S_2) on which is tacked a U-shaped piece of stout wire gauze. The top of the cage is completed by a sliding metal lid.

In the side S_2 are pierced two $\frac{3}{8}''$ holes (H_1 H_2). These may be closed or opened by small metal covers. Over H_1 on the inside of the cage, is fixed a slot of excluder zinc (E). Two metal tubes for candy are inserted into holes H_1 and H_2, the shorter reaching to the excluder in H_1 and projecting $1''$ from the cage, and the longer inserted into H_2 and projecting $2''$ from the cage. Two hinged wires (W_1 W_2) serve to support the cage between brood combs spaced sufficiently wide apart to give a bee-space on each side of the cage.

The queen is inserted into the cage by partial withdrawal of the sliding top, which is then closed. The tubes, filled with stiff candy, are inserted into the side, and the cage suspended in the hive.

Piana states that provided the candy is "hard" the bees will enter the cage via the excluder in about 40 hours and that in another 20 hours the candy in the longer tube will be eaten away and the queen liberated. He advises withdrawal of the cage after 5 days.

4. THE MONT-JOVET CAGE.

In L'Apiculture Française of February 1937, Mont-Jovet describes an excellent cage invented by him (Pl. VII, Fig. 2), embodying the Chantry principle and similar in essential respects to the "Gapir" cage. It is rectangular and made of wire-cloth soldered to tin at the top and bottom and on one side. It can be suspended between two combs when in use.

The candy is contained in two movable tubes (T_1 T_2), one short and one long, inserted through the bottom of the cage. Their openings can be covered or exposed by the movement of a metal slide which is provided with two openings corresponding with the tubes. As distinct from those in the "Gapir" cage the candy tubes are inside instead of outside the cage.

One opening in the slide which admits to the shorter tube is fitted with a piece of queen-excluder (E).

The top face of the cage is provided with an opening through which the queen and her attendants are inserted, or alternatively into which a ripe queen-cell may be partially inserted. The cage therefore serves either for queen introduction or as a nursery cage (p. 176).

Mont-Jovet recommends that the hive be not disturbed for 8 or 10 days after the insertion of a caged fertile queen. By that time she will have escaped and will be laying.

A minor disadvantage in the use of some of the preceding cages is that combs must be spaced apart to admit them. If they are so left for several days the bees will build combs in the space during times of prosperity. This difficulty is avoided by the use of the improved Miller cage (p. 74).

5. THE JAY SMITH "CHANTRY" CAGE.

An ingenious simplification of a cage permitting the use of the "Chantry" principle has been devised by

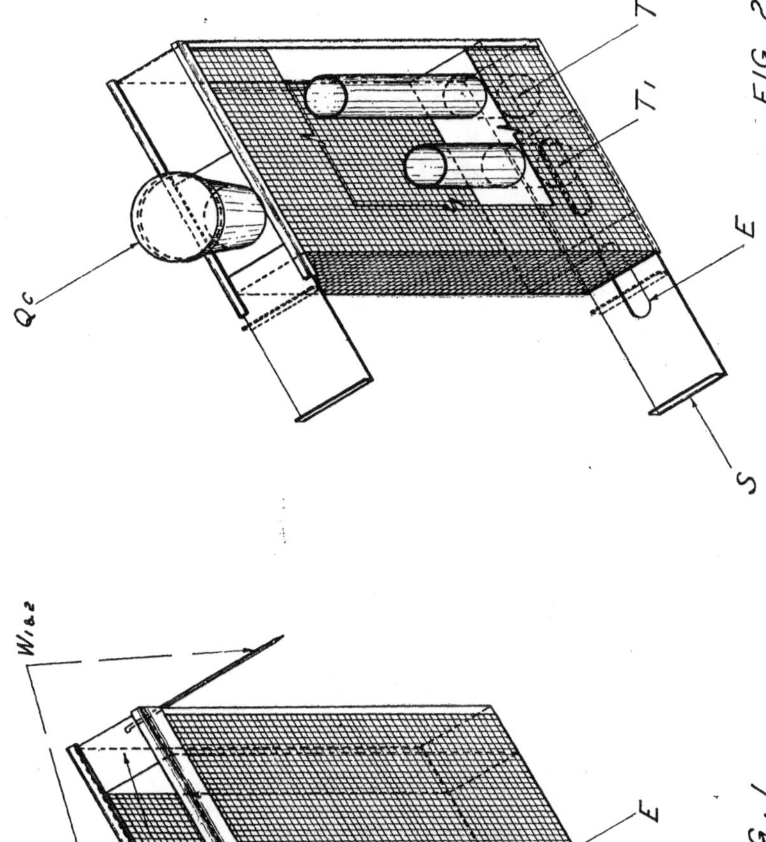

Fig. 2. Mont-Jovet Cage.

Fig. 1. "Gapir" Cage.

Jay Smith.[1] The cage, which he considers to be "the best cage so far we have ever used" is made of $\frac{1}{4}''$ wood, faced with wire cloth, and divided into two compartments by means of a wooden partition. One compartment is for the queen and her attendants and the other for candy, the latter being open at one end where the candy is exposed. The partition extends from this open end to within $\frac{3}{8}''$ of the other end, the total length of the cage being $3\frac{1}{2}''$. At a distance of $2''$ from the open end a hole is made in the partition and over this hole a slot of excluder zinc is fixed.

When in use the hive bees eat away the candy until they come to the hole in the partition. This usually takes about 3 days. The bees can then enter the queen compartment but the queen cannot escape because of the excluder. The bees continue to eat away the candy until they come to the end of the cage—usually after another day. The queen is then able to leave the cage via the $\frac{3}{8}''$ space at the end of the partition.

Jay Smith says "The cage is so sure in its working that we introduce a large number of queens in an out-yard and do not look at the colonies until other work some time later makes a visit necessary".

6. WHYTE INTRODUCING CAGE.

Giraud-Pabou[2] recommends the introduction of a valuable queen to a comb of emerging brood placed between two combs of stores over a sheet of wire cloth placed above the queenless colony.

Gillet-Croix[3] advises this in the case of a queen which is indisposed or fatigued.

---

[1] Smith, Jay, *Bees and Honey*, 1938, p. 273.
[2] Perret-Maisonneuve, A., *L'Apiculture Intensive et l'Elevage des Reines*, p. 282.
[3] Gillet-Croix, A., *Précis d'Apiculture et Sélection des Reines*, p. 51.

Maisonneuve[1] recommends introduction to one comb of emerging brood contained in a wire cloth enclosure, the whole being placed between two combs of brood in the queenless colony.

In England the "Whyte" introducing cage is specially made for this purpose (Pl. VIII). It consists of a metal-bound wire-cloth cage, large enough to enclose completely a British standard frame. The sliding lid (L) is provided with a hole (H) on one side near its centre which can be opened or closed by a movable metal cover (C). Through this hole the queen is introduced.

For use the cage is furnished with a comb of ripe brood from which every bee has been shaken, but from which young bees are emerging. The queen, without attendants, is introduced through the hole in the lid, the cover closed, and the cage then placed in the brood-nest of the queenless stock. The queen finds food and an increasing retinue of young bees in the cage, and soon commences to lay. In about a week's time any queen-cells outside the cage must be broken down. The comb with queen and brood may then be removed from the cage and inserted in the brood-nest.

If a Whyte cage be used in a stock dequeened some days previously, care must be taken that no queen-cells mature whilst the new queen is caged.

Miller[2] introduced valuable queens to several combs of sealed and emerging brood. He described the procedure as follows:—

"If I wish to run no risk whatever, as in the case of a valuable imported queen, I put in a hive without any bees several frames with no unsealed brood, but with plenty of sealed brood, some of it just emerging, and then closing the hive bee-tight put it where there is no danger of the brood being chilled. One

[1] Perret-Maisonneuve, A., *L'Apiculture Intensive et l'Elevage des Reines,* p. 283.
[2] Miller, Dr. C. C., *Fifty Years Among the Bees,* p. 264.

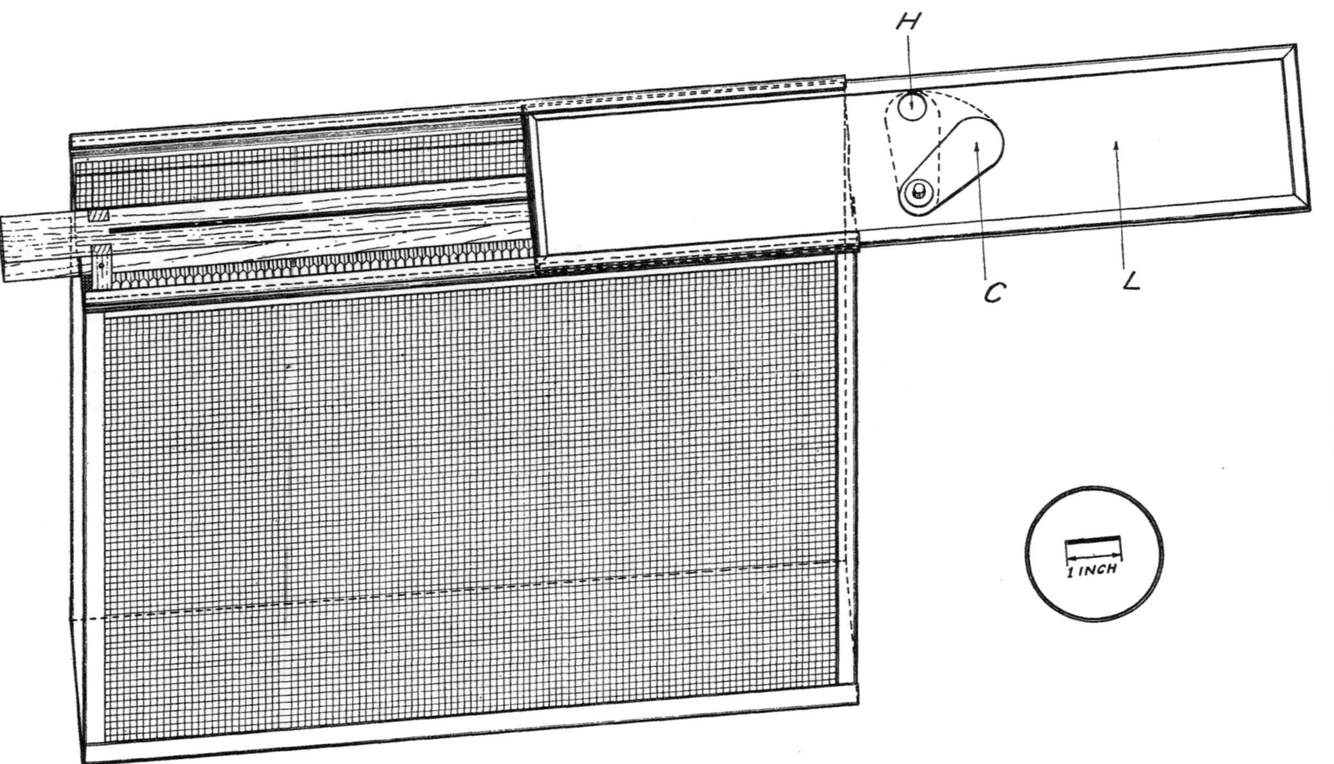

WHYTE INTRODUCING CAGE.

way to do this is to put it over a strong colony, wire cloth preventing the passage of the bees from one hive to the other. At the end of five days the hive can be set on its own stand, and these five-day-old bees, under stress of necessity, will soon be seen carrying in pollen."

Jay Smith[1], commenting on the practice of introducing queens to emerging brood placed over wire screens, rightly points to the danger of a queen being stung, through the screen, by the bees of the stock below.

Arrangements may be made for keeping the combs with emerging brood in a chicken incubator, the temperature of which, according to Meurant[2] should be maintained at from 27° C. to 38° C., that is, from 80° F. to 100° F. It may here be noted that Sladen[3] recommends an incubator temperature of 96° F. to 97° F. for the incubation of queen cells.

INTRODUCTION BY CAGING TO A STOCK CONTAINING FERTILE WORKERS.

Normally there is no worker brood in a stock which has developed fertile workers. The bees, perhaps deluded by the presence of the eggs and larvae of their fertile sisters, probably do not experience that state of complete orphanage which follows the loss of a fertile queen, and are therefore unwilling to accept a new queen introduced to them by ordinary caging methods. If liberated after the usual interval she will be killed— whether by the ordinary bees or by the fertile workers is, the writer believes, not known. It would be natural, one supposes, for the latter to regard her as a rival and to sting her to death.

If however some unsealed worker brood be inserted

[1] Smith, Jay, *Queen Rearing Simplified*, p. 91.
[2] Perret-Maisonneuve, A., *L'Apiculture Intensive et l'Elevage des Reines*, p. 283.
[3] Sladen, F. W., *Queen Rearing in England*, 1905 Edition, p. 29.

into the stock two or three days previously the conditions at once approximate to those obtaining in a normal stock, where a queen is present, or from which she has recently been removed. The bees then need their excess of larval food for the young brood and the fertility of the laying workers tends to decline.

A caged queen may then be introduced, but she should not be released until after an interval of three days.

Herrod-Hempsall[1] recommends that she be caged over unsealed brood, or that alternatively a "Whyte" cage be used (p. 96).

Maisonneuve[2] distinguishes between stocks that have contained fertile workers for a short time and those which have contained them for a considerable time. The former he would requeen with a mature queen-cell after having inserted ripe brood three or four days previously. For the latter he advises shaking the bees at a distance from their home, giving emerging brood, and requeening with a ripe queen-cell or fertile queen. The troublesome operation of shaking the bees however is both unnecessary and unreliable (p. 58).

Attempts to requeen laying worker stocks by caging methods frequently result in failures. The writer has been uniformly successful when introducing by the "One Hour" method even when laying workers have been present for some time. Moreover he has found it sufficient to insert brood of any age twenty-four hours before the queen is introduced.

Hommel[3] recommends the insertion of ripe brood into the hive 8 or 10 days before the introduction of a queen to a laying-worker stock. The resulting young bees will be disposed to accept and protect her. This is doubtless true, but if the queen is available she

[1] Herrod-Hempsall, W., *Beekeeping, New and Old*, Vol. I., p. 697.
[2] Perret-Maisonneuve, A., *L'Apiculture Intensive et l'Elevage des Reines*, pp. 38-138.
[3] Hommel, R., *Apiculture*, p. 435.

should be introduced without delay, the introduction of young brood, or of brood of all ages, a day or two beforehand being all the preparation that is necessary.

INTRODUCTION OF QUEENS TO BEES LONG QUEENLESS.

If during the summer months a stock is kept queenless for some time after it has ceased to have brood, laying workers may or may not appear in it. Potentially fertile workers however are almost certain to develop and for this reason the stock should be regarded as a laying-worker stock for the purpose of queen introduction (pp. 56, 99, 139). If, however, a stock has been queenless during the winter there is no danger of fertile workers and the writer has found that the bees will then readily accept a fertile queen without preliminary insertion of brood.

POINTS TO BE OBSERVED IN CONNECTION WITH TRAVELLING AND INTRODUCING CAGES

1. BEES.

IF a queen is to be caged for a few hours only half a dozen bees will be sufficient for her escort. If confinement is to last for two or three days, or longer, about 15 attendants are desirable, but this number should be augmented to 25 or more, according to the size of the cage, if the weather is cold. Some writers suggest the stocking of a cage with the youngest (newly-emerged) bees, especially when the queen is to be provided with a new escort, mainly because any queen is acceptable to them. These bees however are unsuitable for this purpose because their pharyngeal brood-food secreting glands are undeveloped and they do not therefore feed the queen. Rösch showed that bees from one to two weeks old are the best nurse-bees. Leuenberger[1] states that the pharyngeal glands are most highly developed in bees from 7 to 11 days old and that in very young bees they appear shrunken and out of use. The most suitable bees for caging with a queen therefore are those in their second week, and these, although indistinguishable from rather older bees, are found in greatest numbers on undisturbed combs of young brood. The writer shows in Chap. VIII how a queen can be quickly provided with an escort of strange bees of any age.

Any cage of the Benton type can easily be stocked with bees in the following manner:—

[1] Leuenberger, F., *Les Abilles, Anatomie et Physiologie*, p. 83.

Take a brood comb with adhering bees from the hive and place it on a frame carrier, or stand it on end and leaning against a support. A little smoke over the upper part of the comb will cause the bees to run to the lower part and cluster there whilst many of the older bees fly back to the hive. Place the cage, with wire-cloth removed, under the clustered bees and with a postcard gently scrape into it about the requisite number of bees. Cover the cage immediately with a sheet of celluloid cut a little longer than the cage but of the same width and strap it on with a rubber band. This cover can be moved under the band to open the cage so that any excess of bees may be allowed to escape. Take the cage in the left hand, withdraw the cover about half an inch and place the thumb over the opening so made. The bees keep away from the thumb, which can be safely raised whilst the queen is being put into the cage.

It will now be necessary to substitute the wire-cloth cover for the temporary one. To do this remove the rubber band and keeping the temporary cover in position, place the wire-cloth upon it. Replace the rubber band and withdraw the celluloid from under the wire-cloth.

Alternatively the bees can be placed in the cage one at a time, in which case a temporary cover is not needed. The wire-cloth cover is strapped to the cage by the rubber band, withdrawn half an inch, and the bees inserted under the thumb. Most cages are now provided with a hole in one end for the insertion of bees and queen. This hole is often so small that it is difficult to make the bees, and especially the queen, walk into it. If necessary it should be enlarged with a pocket knife.

With a little practice a cage can be stocked very quickly by inserting the bees one at a time. The bees should be held by both wings so that they cannot sting the operator, and presented to the cage head first. It is easiest to catch those which are drinking honey from the cells and it will be found that they will walk much more readily into a small opening if they are picked up backwards, that is to say, with their heads pointing to the wrist of the operator rather than away from it. This does not apply to the queen. She should be grasped gently but firmly by both wings whilst walking away from rather than towards the hand. If the operator is nervous and makes two or three unsuccessful attempts to grasp the wings, the queen, especially if she is young, will become frightened and may take flight. Should this happen at her hive out of doors she will return to it if the operator stands still and leaves everything undisturbed for a few minutes.

It is always best for an unskilled person to catch or cage a queen indoors before a window. If she then takes flight she can be recaptured. If no building is near, a closed car, though less convenient, will serve.

Many good beekeepers are reluctant to capture queens with the fingers and some are unable to do so. The writer remembers a case in his own experience many years ago. He was examining a candidate for an Expert's Certificate. It was a condition of success that the queen should be caught and caged. The candidate was a middle-aged and very big man with enormous hands. He made several attempts to take the queen by the wings but his fingers were so large and round that he could not possibly do so, and in a last desperate effort he crushed her. It is doubtful whether he could have picked her up by grasping the thorax. Although he was otherwise skilled in bees the writer had no choice but to fail him.

Although it is more expeditious to use the fingers it is possible to carry out all the operations connected with queen introduction without handling the queen.

(a) *To remove the queen from a comb.*

Suspend or stand the comb in a convenient position or get a friend to hold it. Take an ordinary safety-match box, push the drawer three quarters open, and when the queen is in a convenient position, put the open box over her. She will be alarmed and will at once retreat, perhaps with some bees, into the dark part of the box. Gently close this with the finger, whilst moving it slightly from side to side on the face of the comb. Push a drawing pin through the side of the box and label the latter. To isolate the queen open the box on a window-sill and allow its occupants to walk out on to the window. When the queen is alone place the box over her again.

(b) *To take a queen from a cage.*

When a cage containing queen and bees is opened, whether in daylight or artificial light, the queen does not take flight as many people imagine. Both she and the bees are reluctant to leave, but they will do so if disturbed by tapping, smoke, or even the insertion of a finger into the cage. Let her walk out on to a table or window and capture her with a match-box as described above.

(c) *To transfer a queen from a match-box to a cage.*

Open the cage to the extent of half an inch. Place the match-box near the opening and push it open until the opening in the cage is covered by that in the match-box. Strap both together with a rubber band. Cover the match-box to keep it dark but leave the cage in the light. The queen will walk into the cage although she will sometimes take a little time to do so. A little smoke driven against the closed end of the match-box, but not on to the cage, will stimulate her to leave the box, but this is not usually necessary. As soon as she is seen in the cage remove the box and introduce bees to her in the usual way.

2. TEMPERATURE.

Caged bees live comfortably in a temperature of 60° to 70° F. At lower temperatures they may become chilled; at higher ones they become unnecessarily active, try ceaselessly to escape from the cage, and soon soil it with excrement. During warm weather they may conveniently be kept in an ordinary living room, obscured from light. In cool weather they may be placed on a mantel-piece over a room fire. Perhaps the best place in which to keep a caged queen is the pocket —a coat pocket in warm weather and a waistcoat pocket in cool weather.

Queens should not be handled or caged in cold weather. They are very susceptible to cold and quickly die if they are chilled. Mr. Herrod-Hempsall's advice that a hive should not be opened if the temperature is below 60° F. is good. If it is imperative to dequeen a hive in cold or windy weather it should be taken to a sheltered position, or indoors, flying bees being allowed to escape through the doorway.

3. PERIOD OF CONFINEMENT.

Bees should not ordinarily be confined in a cage for more than a week to ten days. If it is necessary to confine a queen longer she should be given a fresh escort (p. 135) and a clean cage.

4. COVERINGS.

Wire-cloth of about 16 meshes to the inch is the best covering for a cage, but perforated zinc of about 12 holes to the inch may be used instead. (Some people think that zinc has an undesirable effect on the queen[1] but the writer has not found any disadvantage in its use.) If the meshes are too close together or the holes too small, external bees may not be able to feed the

---

[1] Hunter, J., *A Manual of Beekeeping*, p. 128.

queen during the process of introduction. On the other hand, if the holes are too large, hostile bees may seize the queen by a foot and even deprive her of a leg.

It should be remembered that cardboard is not a reliable covering for a cage because external bees soon nibble through it. Most cages are provided with openings leading to the candy and these are usually covered by thin card. If one of these cages be exposed to the bees of a hive they may gain premature admission to the candy by tearing the card to pieces. The card should therefore be protected by a piece of tin or perforated zinc until it is intended to admit the bees to the candy.

5. EXIT.

It is important that the exit by which the queen is to escape from a cage be sufficiently large. Some cages are carelessly made and the writer has known of cases where the exit has been obstructed by a small splinter and the queen therefore unable to leave the cage. The opening should be clean and $\frac{3}{8}''$ in diameter.

6. QUILTS.

When an introducing cage is placed under a quilt the latter may settle over the end of the cage so as to obstruct the exit. A match-box or other object placed about an inch from the end of the cage will support the quilt and keep it clear of the end of the cage.

7. FOOD FOR QUEEN CAGES.

The normal food of bees comprises honey and pollen. The former, consisting mainly of water and carbohydrates, provides the bees with heat and energy. The latter provides protein necessary in the formation of muscle and other tissues, and also such vitamins as are essential to healthy animal life. Both honey and pollen provide mineral salts which are essential for the proper functioning of all living organisms.

It is not advisable to provide a cage with honey—even in its candied form—because of its hygroscopic character. It rapidly absorbs moisture from the air, becomes semi-fluid, and causes the death of the bees which become clogged by it. Ordinary candy made from boiled sugar is unsuitable because water of crystallisation may continue to be absorbed by the sugar, with the result that it becomes so hard that the bees cannot eat it. Powdered dry sugar mixed with a very small proportion of honey to make it palatable is used instead. This mixture, first used by Scholtz for feeding stocks of bees and afterwards adapted by Good for queen cages, should be made as follows:—

Take a small spoonful of good liquid honey known to have come from a healthy stock and heat it to make it thin. Pour it into a warm cup and gradually knead into it some "icing" sugar until it has the consistency of stiff putty. If the icing sugar is suspected of containing added starch it is better to use "castor" sugar which should first be ground to fine powder with a pestle and mortar. If the candy is made with cold honey it is apt to become sticky when exposed—as in a hive—to a rise in temperature. This mixture is hygroscopic, but less so than honey. If stored away from air, e.g. in a small screw-capped honey bottle, it can be kept indefinitely and is always ready for use. When placed in a cage it should fill the candy-well and should be covered with water-proof paper or cellophane to prevent contact with the air. The bees eat it slowly and for a time apparently keep in good health on it. As Cheshire[1] says, it contains no material to collect in the bowel.

"Good's" candy however cannot be a complete food for bees, being almost[2] entirely devoid of protein,

---

[1] Cheshire, F. R., *Bees and Beekeeping*, Vol. II, p. 350.
[2] Honey contains minute and variable proportions of pollen grains.

vitamins, and mineral salts. That bees kept on it do not display signs of dysentery is sometimes taken to prove that it is an ideal cage food. Yet when bees and queen are fed on it for more than two or three days—(two or three weeks in some cases of queen importation)—there must be protein deficiency accompanied by a serious lack of vitamins. May this not well be the cause of the reputed inferiority of queens which have travelled a long journey by post?

Caged bees, however, having little scope for activity and no brood to feed, need relatively little nitrogenous food, but if caged for more than a day or so they must need some. Some writers recommend the addition to the candy of a little wheat flour as in the case of "Viallon" candy,[1] but ordinary white flour contains 72 per cent. of starch, which, as Phillips[2] has shown, is entirely unavailable as food for bees. In experiments by the writer and others on pollen substitutes, cereal and pea flour have been shown not only to be useless but inhibitory to the production of brood. They cannot therefore be suitable for mixing with candy for queen cages.

Bees caged with sugar candy only may for a short time feed a fertile queen with perfect secreted food at the expense of their own bodies, but if their confinement is prolonged nitrogen deficiency is inevitable. In such cases virgin queens fare worse for they are not fed by the bees.

In the writer's experiment described on p. 151, in which six queens with attendant bees were confined in one cage for three months, two or three cells-full of fresh pollen were placed in one part of the cage at weekly intervals. The bees ate it slowly—doubtless as they needed it—and showed no signs of dysentery or apparent congestion of the colon throughout the period. This

---

[1] Cheshire, F. R., *Bees and Beekeeping*, Vol. II, p. 337.
[2] Snodgrass, R. E., *Anatomy and Physiology of the Honey Bee*, p. 176.

pollen was not mixed with the candy or the bees may have eaten too much.

For ordinary queen-cages therefore the writer recommends the admixture of a very small amount of fresh pollen to "Good's" candy—one cell-full to about as much candy as would be contained in an egg-cup. This makes a more complete food and ensures that the bees and queen are reasonably well nourished.

The rate at which candy is consumed by caged bees depends on the number of bees, the consistency of the candy, and temperature. Care must be taken that it is replenished as necessary.

As a rule the bees of a good stock will eat their way through the well of stiff candy in a Benton type cage in about $1\frac{1}{2}$ days. If the candy be guarded by a sheet of rough brown paper or very thin card, pierced with pin holes, they will be delayed by an additional half day. In the case of cages embodying the Chantry principle the larger column of candy must be of considerable length because the bees consume it at both ends.

The candy-well and the tunnel leading to it should be filled with stiff candy. If this has become soft owing to exposure to air, more sugar should be kneaded into it. Some writers recommend that the candy-well be coated with wax—varnish would be suitable—to prevent absorption of moisture by the wood. This is unnecessary when "Good's" candy is used because this tends to become moist, but is desirable if boiled candy be used.

In hot weather caged bees, especially those arriving through the post, may need water. This may be provided by wetting the fingers with water and rubbing them gently over the perforated zinc or wire-cloth covering. Water should not·be allowed to drop into the cage.

8. CLEANLINESS.

Cages should be scalded and scrubbed after use and stored in a dry place. They should not be provided with candy until immediately before use.

9. PACKING FOR POST.

A cage containing queen and bees should be suitably packed before being consigned to the post. Although post officials take reasonable care and (in Great Britain) go so far as to ensure that suitably labelled cages are not transferred from a train by mechanical apparatus, some violent shocks for the bees are almost inevitable. It is usual to cover the wire-cloth surface of a cage with cardboard or thin wood, allowing for ventilation by the pierced side grooves and the wire-cloth covered open end. In hot weather the ventilation is sometimes insufficient, especially when a cage is buried in a mass of letters. The writer's method of packing, which is simple and has always proved satisfactory, is as follows:—

The wire-cloth surface is not covered by cardboard or wood. A piece of new corrugated paper is cut 3″ longer than the cage and sufficiently wide to be folded round it so that it is double only on the top surface. The corrugations are on the inside next to the cage and parallel to its length. The overlapping portion of the paper is pasted to that beneath it, care being taken not to crush the corrugations. String is then tied tightly round the middle and crossed to go over the ends through holes pierced in the paper close to the cage. This leaves the package open at both ends and allows air to pass through the tunnels in the paper to the wire-cloth covering of the cage. Similarly air passes along to the ventilating side grooves. There is some resilience in the projecting ends of the paper which mitigates the violence of any blows which may be received in transit. A small adhesive label bearing the names and addresses of sender and recipient is placed on the package itself, and a "tie-on" label attached. This bears the warning "FRAGILE. LIVE BEES. WITH GREAT CARE," the name and address of the recipient, and at its remote end, the postage stamp. A little red sealing-wax appropriately applied to the package draws attention to its importance and ensures considerate handling.

10. RECEPTION OF ESCORT.

It is usual, and in accordance with instructions accompanying travelling cages, to introduce the queen, together with the bees with which she has travelled, to a queenless colony. There can be little doubt that mutual unfriendliness is largely due to the presence of the escort bees in the cage and that this is to some extent unfavourable to the reception of the queen when she is liberated. Careful observation will show that in many cases the escort bees are killed and thrown out of the hive after liberation whilst the queen is favourably received and accepted. This can be proved by marking the bees of the cage and is often noticeable when Italian queens with bees are introduced to stocks of black bees.

In Chapter VIII reference is made to the advantage of giving a travelled queen a new escort from the hive into which she is to be introduced. If for any reason this cannot be done it is better to allow the attendants to leave the cage and to introduce the queen alone.

CHAPTER VII

DIRECT INTRODUCTION

ADVANTAGES OF DIRECT INTRODUCTION.

Economy of time and labour is important to most beekeepers, especially to those who keep their bees in out-apiaries. The introduction of a queen by an ordinary caging method usually involves three or four visits to the hive, viz:—

(1) To dequeen.
(2) To insert the cage 24 hours later.
(3) To arrange for the liberation of the queen after a further 24 hours.
(4) To remove the cage and to ascertain whether the queen has been accepted.

In a case of unfavourable reception an additional visit is necessary between (3) and (4). It is possible of course to dequeen the hive and insert the cage at one time and to arrange for the queen's release after 48 hours. The time necessary for preparing, stocking, and provisioning a cage is considerable and the need for returning to the hive at prescribed times is often inconvenient.

Introduction by caging may result in a loss to a hive of several thousands of bees, for a caged queen is not laying during her imprisonment nor does she usually begin to do so immediately after her release.

A queen from a stock infected with Foul Brood may be safely placed at the head of a healthy stock by direct introduction. This is stated by Hutchinson[1]

---

[1] Hutchinson, W. Z., *Advanced Bee Culture*, p. 98.

and has been confirmed by experiments of the writer. The same queen introduced with her attendants by a caging method would be likely to transmit the disease.

Apart from these considerations there is no more safety in caging methods than in the best methods of direct introduction. The latter involve less visits, less work, and less disturbance; no breaking down of queen cells, and no subsequent adjustment of combs (as in the case of "press into comb" cages). Moreover the laying of the queen is interrupted for a few hours only instead of a few days.

Even the use of the best cages, i.e. those embodying the Chantry principle, necessitates two visits to a hive, one to insert and the other to remove the cage, and if the combs are widely spaced to admit of the insertion of these cages the space so made tempts the bees to build fixed combs in it.

The best known methods of direct introduction will now be described. In appraising their respective merits it is necessary to consider the time, labour, and number of visits involved, as well as the probabilities of success. Concise instructions for using them are given in Chap. XII.

It must be premised that in all cases the conditions of stocks to which queens are to be introduced are such as are favourable to their reception (v. Chap. III).

1. METHOD OF IMMEDIATE EXCHANGE.

This is perhaps the simplest of all methods of introduction, but it involves considerable risk. The old queen is found and removed with the minimum of disturbance to the stock, and the new queen is immediately put in her place on the comb where she was found. The comb is replaced in the hive without jarring or noise and the hive covered. The bees are unaware of the change if the new queen is "in full lay". This method should be used with the greatest caution. It is not always possible to find the old queen without disturbance and it is essential that both queens be laying and that the stock be prosperous as during a period of honey flow. Ruffy, to whom Maisonneuve[1] attributed this method, advises that the comb of bees with the new queen be lightly powdered with flour before replacement in the hive. It is hardly worth while to incur the risks involved in this procedure when the Water Method (p. 130), which is equally rapid and most reliable, can be used instead.

2. "CHATELAIN" METHOD.

This is described by Halleux[2] who gives the following directions:—

Remove the old queen and any queen cells which may be found in the hive. (It may here be stated that it would be very unwise to introduce a new queen if the hive previously contained both a queen and queen cells, for the new queen would soon issue with a swarm.) Wait until about three-quarters of the bees show distress at their loss, i.e. after five or six hours, and then present the caged queen to the distressed bees at the hive entrance. If these crowd round the captive queen, beating their wings and offering her food, release her and let her run in at the entrance. Maisonneuve,[3] who recommends presenting the queen to the bees on the combs instead of at the hive entrance, wisely remarks that acceptance may be localised and that the queen may subsequently meet hostile bees in another part of the hive.

Presumably the queen is to be introduced by another method if the bees do not show friendliness to her.

[1] Perret-Maisonneuve, A., *L'Apiculture Intensive et l'Elevage des Reines*, p. 281.
[2] Halleux, D., *Le Livre de l'Apiculteur Belge*, p. 366.
[3] Perret-Maisonneuve, A., *L'Apiculture Intensive et l'Elevage des Reines*, p. 284.

This method, necessitating two visits to the hive and involving much uncertainty, cannot be recommended.

A writer in the "Australasian Beekeeper" of September 1938 describes a variation of this procedure. The hive must be queenless and without queen cells. The queen to be introduced is caged alone for from 20 to 30 minutes. The cage is then presented to the bees at the entrance of the hive. If they "buzz happily and show no sign of balling" the cage is opened and the queen walks out and into the hive. If, on the contrary, the queen shows fear and races about in the cage, and if the bees cling closely to the cage and show signs of a desire to sting and ball her she must not be liberated, and a further search for queen-cells must be made.

This method has the negative virtue, as the above-mentioned writer observes, "that we know at once, without any risk of losing the queen, if she will not be accepted".

3. Shaken Queenless Swarm Method.

If bees are deprived of their queen, brood, and food, and confined in a ventilated box or hive for a few hours they will invariably accept any fertile or virgin queen which may be dropped amongst them.

The queenless bees should be shaken from their combs into the box, which should be closed with cloth or perforated zinc, and kept in a dark place for from four to six hours. The bees will be anxious to escape and will usually continue in a state of commotion until a queen is given to them. She is simply inserted through a convenient hole in the top or side of the box, the hole being corked when not needed for this purpose. The box is left in the dark for another hour or two, after which the bees and queen will have clustered as a swarm which may be hived on combs in the usual way. It is best to give brood-less combs at first but if a little brood is given it should be sealed so that the bees may not be tempted to raise queen cells.

This is an excellent method of introducing virgin queens (p. 170) and is much used in the stocking of nuclei used for mating. The mating boxes are stocked with sufficient bees about mid-day, closed, with adequate ventilation, and kept in a cool dark place until evening. Virgin queens are then dropped into them and an hour or so later the necessary combs are inserted (p. 170).

Doolittle[1] stocked nucleus hives by this method. He used bees from supers placed over excluders so as to avoid the possibility of including the queen. He shook them into the nucleus boxes through a rectangular funnel of convenient size and confined them for 7 or 8 hours. Before introducing a queen he set the box on the ground with a little violence so as to break up the cluster and throw the bees to the bottom of the box. This ensured that the queen would fall amongst the bees and that they would be too terrified to molest her.

Doolittle writes that the confined bees literally cry for a queen after three or four hours but that they may "ball" her if she is introduced too soon.

4. RAYNOR'S SHAKEN BEES METHOD.

Raynor[2] used the shaken bees method for requeening a skep. He drove all the bees from their home, in the usual way, into an empty skep, removing their queen as he did so. He then replaced the original skep on its stand and threw the driven bees on to a cloth in front of it. He dropped the new queen (with her escort

---

[1] Doolittle, G. M., *Queen Rearing*, p. 60.
[2] Raynor, G., *Queen Introduction. The Ligurian Bee.* A paper read to the B.B.K. Association, January, 1880, p. 7.

bees, if any) amongst the terrified bees as they ran back into their home.

The same method is applicable to a frame hive and although it involves considerable labour, is reliable. Raynor states that he used it hundreds of times without a single failure, and Doolittle[1] speaks of a similar plan as infallible for introducing a valuable queen.

The distinctive feature of this method is that the bees, deprived of their queen, are first driven into an empty hive before they are returned to their own home.

5. PARTLY SHAKEN BEES METHOD.

Some writers recommend the shaking of only a part of the queenless stock. Two or three combs, well covered with bees, are taken from the centre of the queenless hive and the bees are shaken or brushed from them on to the alighting board. As they run into the hive the queen is dropped amongst them. Hommel[2] recommends that both bees and queen be sprayed with sweetened scented water as they run in. This method he describes as "nearly certain". It is, of course, better to shake all the bees of the stock although this entails considerable labour.

Introduction by this method is effected in one operation and is usually successful, but the labour involved, together with some uncertainty, makes it less desirable than several other direct methods. It must always be remembered that shaking bees from combs during a honey flow results in the bees being half drowned in the nectar which falls from the cells, and shaking in times of scarcity may lead to robbing.

6. NUCLEUS METHOD.

By this method a nucleus stock, containing a young fertile queen, is simply placed in the midst of a queen-

<hr>

[1] Doolittle, G. M., *Queen Rearing*, p. 81.
[2] Hommel, R., *Apiculture*, p. 433.

less colony. This must be done however with due caution. Essential conditions for success are that the colony must be prosperous and that there be sufficient bees with the queen to tend and protect her whilst the bees of the stock are becoming acquainted with her.

Simmins[1] introduced this method in 1881. He inserted into the queenless colony one comb of bees with queen but took the precaution of exposing the comb to light whilst carrying it from the nucleus hive to the stock. In the writer's view it was rather exposure to air which was beneficial as diminishing the distinctive odour of the bees. Simmins considered that the method was successful in all circumstances and at any time of the year but the writer would make some reservations having experienced failures during times of scarcity.

Hutchinson[2] says that when a colony has been queenless long enough to build a batch of queen cells, he takes a comb with adhering bees and queen from a nucleus hive and hangs it in the colony near one side of the hive from which he has previously driven the bees with a little smoke or a feather. The side of the comb on which the queen is found is turned towards the wall of the hive so that the bees of the stock make her acquaintance gradually. Finding the queen attending to her duties and surrounded by a cortege of her own bees they do not molest her. It may here be observed that bees building queen cells would be expecting a queen and therefore would be the more favourably disposed towards the stranger. There would be some danger to a queen introduced in this way to a stock recently dequeened. In such a case Maisonneuve[3] recommends the removal of all unsealed brood

[1] Simmins, S., *A Modern Bee Farm*, p. 282.
[2] Hutchinson, W. Z., *Advanced Bee Culture*, p. 94.
[3] Perret-Maisonneuve, A., *L'Apiculture Intensive et l'Elevage des Reines*, p. 284.

—(this deprives the bees of the hope of raising queen cells)—and leaving the hive queenless until the following day. Two combs of bees with a queen, taken from a nucleus, are then inserted in the centre of the brood-nest which has been previously subjected to smoke. This plan entails too much labour and therefore is not to be recommended.

Doolittle[1] requeened hives by the nucleus method in the following way:—

He removed the comb of brood with the queen, together with another comb, from the nucleus hive. He carried these to the hive to be requeened, placing them aside, with the queen between the two combs, whilst he searched for and disposed of the old queen. During this interval the bees from the nucleus had clustered between the two combs. He then removed two combs, with bees, from the hive, and inserted the two nucleus combs together in their place.

It will be observed that the bees of both nucleus and stock were exposed for some time to the air.

The Rev. Bro. Adam[2] of the famous Buckfastleigh Abbey apiary recommends the removal of three combs with bees and the insertion in their place of a 3-comb nucleus with queen, during the honey season, i.e. in early July, and that the stock be thereafter left uncovered for five minutes. He states that this plan has given eminently satisfactory results and considers that it is absolutely safe.

Sellner[3] recommends uniting a nucleus containing a young mated queen with a queenless stock in the following way:—

Place the nucleus, without a floor-board, over perforated zinc, over the brood-nest. Leave it for at least 24 hours and on the evening of the following day

---

[1] Doolittle, G. M., *Queen Rearing*, p. 76.
[2] Adam, Rev. Bro., Article on Queen Introduction, Somerset Beekeepers' Annual Report, 1934, p. 19.
[3] Sellner, Karl, *Natürliche Königinnenzucht*, Bienen Vater, 1937, p. 198.

quietly withdraw the perforated zinc screen by means of a wire previously attached to it, without disturbing the bees. Examine the hive on the fifth day afterwards, when the queen should be found to be laying.

The writer has often requeened a stock by means of a nucleus:—

(*a*) By placing a strong nucleus on the hive stand to receive the field bees—this only during a honey flow—and uniting with the parent stock later.

(*b*) By inserting a nucleus into the back of the queenless stock behind a loosely fitting division board. This should be done at dusk and without disturbance of the main stock. The bees mingle gradually during the night and the queen is accepted.

In all cases of nucleus introduction it is desirable that the nucleus be as strong as possible. It is a good thing to strengthen it beforehand with one or two combs of emerging brood from the colony to be requeened. The comb carrying the new queen should, if possible, be placed near the hive wall with the queen remote from the main stock. She should be undisturbed and quietly parading on the comb so that she is not likely quickly to run amongst the strange bees.

A queen-right nucleus should not be inserted into a queenless stock that is without stores. Even if the nucleus were well provided with food, fighting would be likely to ensue and the queen and many bees would be killed. The nucleus method must be regarded as rather uncertain except during a honey flow and the cautious beekeeper would prefer to strengthen the nucleus with brood from the stock to be requeened and then to unite both by the newspaper method (p. 124).

7. Smoke Method.

Introduction with the aid of tobacco smoke was first practised by the American queen breeder Henry Alley.[1]

Doolittle[2] describes how he introduced queens with the aid of smoke. After dequeening a hive he blew smoke into the entrance and at the same time pounded the hive with his fist until the bees roared with alarm. He then let the new queen run in at the entrance, following her up with a puff of smoke. If there were danger of robbing he deferred introduction until nightfall.

A. C. Miller[3] improved on this rather "rough and ready" procedure and gave more precise instructions. He reduced the hive entrance to a width of about one inch by closing the remainder with grass, cloth, or wood. He then blew into the hive three or more strong puffs of thick white smoke and at once closed the entrance completely. The amount of smoke necessary would vary according to the size of the hive. In any case it should be sufficient to make the bees roar with fright within fifteen to twenty seconds. After this interval he opened the small space at the entrance and allowed the queen to enter the hive followed by a slight puff of smoke. Both bees and queen were thus thoroughly alarmed and concerned only with their safety. The small entrance was immediately closed again and left for ten minutes. It was then reopened and the bees allowed gradually to cool down and to obtain ventilation. The full entrance was restored an hour or more later or preferably on the following day.

Fernes[4] who relates that he has used the smoke method "under every conceivable condition", reduces

---

[1] *Gleanings in Bee Culture*, June, 1936, p. 336.
[2] Doolittle, G. M., *Queen Rearing*, p. 76.
[3] Perret-Maisonneuve, A., *L'Apiculture Intensive et l'Elevage des Reines*, p. 279.
[4] *Gleanings in Bee Culture*, June, 1936, p. 336.

the hive entrance to about 4 inches in width, and into this he injects "lots of smoke" until the bees roar. He then uncovers the hive and drops the queen between the combs. He re-covers the hive immediately lest the queen "may take fright at the confusion and smoke and fly away". When the bees have become quiet he restores the full entrance.

Owing to the uncertainty of the smoke reaching all the bees Root[1] advises that this method be not practised on a hive comprising more than one story or in one which is only partially occupied by bees. Maisonneuve[2] calls attention to the danger of stifling the bees in hot weather and Pellet[3] cautions the novice against placing too much confidence in the method on account of many reported failures on the part of experienced beekeepers. Like others, the method would be most likely to be successful if carried out in times of prosperity and late in the evening when all the bees are at home. Since it savours somewhat of cruelty however most beekeepers would prefer to use other and easier methods.

The unreliability of the "Smoke" method is indicated by Keen[4] who concludes a description of it by recommending an examination of the hive after an interval of five minutes, and adding, "If there is a ball of bees and queen then you have failed to give them enough smoke".

8. PAPER METHOD.

If it is desired to introduce a queen from a nucleus and at the same time to strengthen the stock by the brood and bees of the nucleus without any risk this can be done as follows:—

[1] Root, A. J. and E. R., *ABC and XYZ of Bee Culture*, 1929, p. 469.
[2] Perret-Maisonneuve, A., *L'Apiculture Intensive et l'Elevage des Reines*, p. 280.
[3] Pellet, F. C., *Productive Bee-keeping*, p. 116.
[4] Keen, G. H., *Introducing Queen Bees*, p. 8.

Transfer the nucleus to a brood chamber and strengthen it by as much sealed brood taken from the stock to be requeened as it can care for. The added combs will tend to give the nucleus the same odour as the stock. When the nucleus is strong dequeen the stock and unite the nucleus to it by the newspaper method as follows:—

Towards night remove the quilts from the queenless stock and cover it with a single unbroken sheet of newspaper. A few holes may be pierced in the paper with a match-stick. The holes must not be large enough to allow bees to pass through them. Place the brood-box containing the nucleus on the newspaper and cover it with the quilts. During the night the odour of the two stocks will become the same, the bees will nibble their way through the paper and unite gradually and peaceably.

If the union be timed so as to take place during the day there may be some fighting, with consequent danger to the queen. It should therefore always take place during the night.

If the nucleus is fairly strong the queenless stock may be placed over it when uniting with newspaper.

This is a perfectly safe method of introduction but there is the disadvantage of using two brood boxes and the necessity of subsequently reducing them to one. Union by the newspaper method should not be attempted if the nucleus or stock is very weak or if the bees have clustered for winter owing to cold weather. In such cases the bees may not tear away the paper before the nucleus dies of cold or starvation.

9. SIMMINS "FASTING" METHOD.

This well-known method is both easy and generally satisfactory. Simmins[1] states that "it meets all requirements whether the colony has been long or only a

____

[1] Simmins, S., *A Modern Bee Farm*, p. 283.

short time queenless; if it has brood or not, or queen-cells in any stage of development. It is also applicable to any season of the year". The directions as given by Mr. Simmins are as follows:—

(1) Keep the queen quite alone for not less than 30 minutes.
(2) She is to be without food meanwhile.
(3) She is to be allowed to run down from the top of the frames after darkness has set in, by lamp-light.

The hive is not to be opened for 48 hours.

Cheshire[1] reports unqualified success with this method in dozens of experiments, including cases where stocks had been queenless for 10 days, 4 hours, and 30 minutes respectively. Root[2] does not recommend this method in the case of a valuable queen, or in any case to a beginner. The writer used it for several seasons many years ago. One or two failures he attributed to the fact that the bees had not been sufficiently long queenless. There were no failures when the hives were dequeened at mid-day and the introduction carried out the same night. Simmins' instructions may be amplified as follows:—

(1) Dequeen the hive about 6 hours before introduction.
(2) Keep the queen in a match-box in a vest pocket during the 30 minutes of her confinement.
(3) Give a light puff of smoke to drive the bees down before introducing the queen.
(4) Push the match-box open over the combs, and cover immediately without noise.

This method depends on the distress of the bees at the loss of their queen, the hunger of the queen when

[1] Cheshire, F. R., *Bees and Beekeeping*, Vol. II, p. 345.
[2] Root, A. J. and E. R., *ABC and XYZ of Bee Culture*, 1929, p. 471.

liberated, and the absence of disturbing influences when she is introduced. It is excellent in respect of reliability but it involves two visits to an apiary as well as working after darkness has set in.

10. HONEY METHOD.

This crude and unattractive method is described by several writers. It consists in immersing the queen in a little warm honey, rescuing her with a tea-spoon, and dropping her through the feed-hole of the queen-less stock between two combs which have previously been brought near enough together to prevent her from falling to the floor. The bees cannot recognise her as a stranger and when they have licked her clean they usually accept her. Cheshire[1] observes that this method is reliable during a honey flow, but most uncertain as well as unsuitable during early spring or late autumn. Doolittle[2] found that it was necessary to keep a stock queenless for from three to five days before introducing a queen in this way. Herrod-Hempsall[3] advises the use of honey from the stock to which the queen is to be introduced as this assists in imparting the hive odour to the queen. Alley[4] who first proposed this method used honey diluted with water. Others have used sugar syrup. There are serious objections however to the use of these substances for they temporarily close the queen's spiracles so that she cannot breathe and she has to submit to the prolonged and probably irritating attentions of the bees which clean her. At the best the method must be considered as a "rough and ready" one which would not be used by a considerate beekeeper.

---

[1] Cheshire, F. R., *Bees and Beekeeping*, Vol. II, p. 344.
[2] Doolittle, G. M., *Queen Rearing*, p. 77.
[3] Herrod-Hempsall, W., *Beekeeping, Old and New*, Vol. I, p. 680.
[4] Perret-Maisonneuve, A., *L'Apiculture Intensive et l'Elevage des Reines*, p. 272.

11. Flour Method.

Before the advent of the newspaper method of uniting bees it was common to recommend the use of flour for this purpose. The bees on the combs of two stocks to be united were thoroughly powdered with flour from a dredger and the combs alternated and placed in one hive. The bees were so incommoded that they disregarded the presence of strangers and acquired a common odour whilst divesting themselves of the troublesome powder.

Flour has been similarly used for the introduction of queens. Under circumstances most favourable to reception, e.g. during a honey flow, and at night, it is sufficient to roll the queen in flour and place her on the combs with the minimum of disturbance. Under less suitable conditions failures may be expected. Pellett[1] recommends smoking the stock (as in the Smoke Method), dropping the queen into a dish of flour and letting her run into the hive. Bertrand[2] recommends shaking of all the bees on to a cloth in front of the hive, sprinkling them with flour, and dropping the queen—also floured—amongst them as they run in.

It will be noted that these writers suggest the use of flour as an additional safeguard in the employment of the "Smoke" and "Shaken bees" methods. Dusting a queen with flour is less objectionable than daubing her with honey but at the same time it is not a desirable or indeed a reliable means of introduction.

12. Chloroform Method.

Under the influence of an anaesthetic bees are naturally unable to resent the intrusion of a strange queen and when they recover from its effects they accept her without challenge. At one time the fumes

---

[1] Pellett, F. C., *Productive Bee-keeping*, p. 118.
[2] Bertrand, E., *La Conduite du Rucher*, p. 40.

of burning "puff-ball", a fungus of the genus Lycoperdon, was used either to stupefy bees whilst their honey was being taken, or whilst a queen was being introduced. Chloroform was found to be equally effective and less unpleasant to use and A. D. Jones[1] of Canada first formulated directions for its use, which, according to Cheshire and Maisonneuve, are as follow:—

Place a small dry sponge in the bottom of the smoker and upon it place another sponge impregnated with a teaspoonful of chloroform. Upon this place another dry sponge and close the smoker. Go to the queenless hive and puff the chloroform fumes strongly into the entrance for a quarter of a minute. Wait two minutes, give a few more puffs, and allow the queen to run in. If this is done during the day whilst the bees are flying give a few more puffs two minutes later.

The purposes of the three sponges are obvious. The first serves to raise the impregnated one to the level of the air inlet of the smoker and the third to prevent the premature dissipation of the chloroform. The writer, however, who has used this method several times with complete success finds that it is sufficient to use a loosely made cartridge of corrugated paper and to pour the spoonful of chloroform on to this. Chloroform evaporates rapidly and it is necessary to use it quickly.

An alternative method, used by the writer, is to take two pieces of corrugated paper, about 4 inches square, pour about two-thirds of a teaspoonful of chloroform on each; quickly insert one under the quilts at the back of the hive and the other into the entrance at the front. Close the latter with a duster for three minutes. Withdraw the papers and drop the queen into the hive at the feed-hole. Cover the hive and open wide the entrance. This should be done in the evening when all the bees

---

[1] Cheshire, F. R., *Bees and Beekeeping*, Vol. II, p. 340.

are at home because otherwise the drowsy bees would be at the mercy of robbers for a considerable time.

If this method be carried out in an observatory hive the bees will at first be seen running over the combs in their efforts to escape the fumes. They will then become quiet and many will drop to the floor as though dead. This alarms the observer, but an hour later every bee will have recovered and the queen, if fertile, will resume her laying as though nothing had happened. The writer has found a full comb of eggs from the newly introduced queen twenty-four hours after introduction although of course a hive so requeened should not ordinarily be disturbed for several days. He has also succeeded in introducing an elderly virgin queen to replace a fertile one by this method. The virgin became fertile within a week. It is not likely of course that this would often be practicable, and it may be that in this particular case the young queen had been previously mated. Jones[1] reported having successfully introduced 50 queens in 50 minutes by this method and to have succeeded in the most difficult cases, including those of fertile worker stocks.

The method must be considered a good one. It is hardly necessary to say that there are some dangers in the use of chloroform and that consequently it is not easily obtainable by people inexperienced in its use.

Cotton[2] 1842, referring to the harmless effects of anaesthetising bees by means of the puff-ball, says:—

"The fungus does them no harm, it only makes them drunk, which is very good for Bees, though bad for men, as they get well in twenty minutes, have no headache next morning, and are all merrier afterwards, and it was not their fault that they were so overtaken."

[1] Cheshire, F. R., *Bees and Beekeeping*, Vol. II, p. 340.
[2] Cotton, W. C., *My Bee Book*, p. 66.

I

13. Snelgrove "Water" Method.

This is the easiest and most expeditious way of introducing a fertile queen to a queenless stock. Its chief advantage is that it is effected in a few minutes and necessitates only one visit to a hive. An account of its origin may be interesting. Sometime in the year 1910 the writer was visiting a small out-apiary shared by himself and a friend. One of the hives happened to be queenless. The friend, who was requeening one of his, suggested that his deposed old queen might go into the queenless hive. No cage was available and the writer, knowing of no method of introduction which could be carried out without a further visit to the apiary, and unconcerned as to the safety of the queen, gave her a bath of saliva in the palm of the hand and dropped her through the feed-hole of the quilt amongst the combs of the queenless stock. At the next visit to the apiary she was found to be laying. The transition from saliva to water was natural. Several experiments in the same year proved successful and in the following year the writer demonstrated the efficacy of the method to a large gathering of beekeepers, including the late Mr. T. W. Cowan, at Cannington. The queens of two stocks were exchanged in a few minutes and an hour later both were shown to be quite safe and parading the combs as if no change had been made. Both were subsequently reported by the local expert as doing well. Since that time the method has become well-known and is practised by many experienced bee-keepers to whom economy of time is important.

The procedure is as follows:—

Dequeen the stock. Take the new queen by the wings and dip her into a cup of luke-warm water, moving her to and fro a few times to ensure that she is thoroughly wet. Remove her, and immediately place her amongst the bees on a comb, or preferably put her through the feed-hole on to a top bar. Cover the hive immediately and without disturbance, and do not examine it for several days.

Alternatively:—

Dequeen the hive. Place the new queen in a match-box. Immerse the box in tepid water, opening it slightly until the water nearly fills it. Close it, remove it from the water, and agitate it gently for a few seconds. Push it open over the feed-hole of the stock so that the queen can walk down between the combs. Cover the hive quietly and do not examine for several days.

It is an improvement if the queen be kept without food for a few minutes before immersion. She will then be hungry and solicit food when she recovers from her bath. The operator may use his discretion as to the temperature of the water. In very hot weather it needs scarcely to be warmed, and may even be used cold. At other times its temperature should be somewhere between 90° F. and 100° F. Experiments have shown that the latter temperature is very suitable. Once a beekeeper has tested water at this temperature with his fingers he will not need to use a thermometer.

A queen suffers no harm from a warm bath of 5 to 8 seconds. She dries in a few minutes after introduction and resumes laying on the same day.

This method can be used at any time of the day and at any period of queenlessness. Queen-cells, if present, should previously be destroyed, and the introduction should be effected, if convenient, in the evening.

The method is especially suitable for substituting one laying queen for another. If bees have been long queenless some brood should be inserted into the hive a day or two before the queen is introduced.

In the case of travelled queens the precautions indicated on page 40 should be observed.

The water method is specially serviceable when it is desired to provide a queen with a new escort. The queen may be immersed and introduced to a cage of bees, or, if the operator prefers not to handle the queen, the bees may be immersed one at a time and introduced to the queen. When a new set of attendants has been provided it is as well to prevent the bees from having access to candy for half an hour and it is an advantage to keep the cage in the darkness of the pocket during this period.

14. Snelgrove Board Method.

When a beekeeper wishes to introduce a specially valuable queen in the safest possible way he should introduce her to young bees only (p. 47). This is easily done by means of a Snelgrove Swarm Control Board[1] as follows:—

Shake the bees from several combs of the ripest brood and put the latter into an empty brood box. Put some empty combs in their place in the original brood box and cover them with an excluder. Over this place the box of shaken brood combs. Leave the hive for half a day or more. The bees (mainly young nurse bees) will pass through the excluder and occupy the combs in the upper box.

Insert the board in place of the excluder and give one entrance in the board to the upper stock. After one day of good flying weather all the old bees will have left the upper stock and have rejoined the queen below the board. Young bees only will be left in the upper stock and to these a queen can be safely introduced by any good method. Both queens can be left to develop good brood nests before the old queen is removed and the stocks united. The union can be effected by means of newspaper (p. 124) or by simply removing the wire cloth over the centre hole of the board, which, of course, should be removed altogether before the winter. The newspaper method is the safer.

[1] Snelgrove, L. E., *Swarming, Its Control and Prevention*, p. 24.

The writer's attention was drawn to this special use of the board by Capt. M. Paine of Sidmouth.

15. Introduction to a Nucleus of Young Bees.

A nucleus containing only young bees may be formed in any stock with a single brood chamber by the insertion of a closely-fitting division board.

A hole about 3″ square should be cut in this division-board and covered with perforated zinc. The board is inserted between the combs so that a nucleus of three combs containing brood, stores, and bees may be formed at the back or side of the hive according to the direction of the combs. A new entrance at the back or side of the hive is made for the nucleus. The main part of the stock with the old queen occupies the compartment on the other side of the division board and continues to use the original hive entrance.

When the nucleus is formed one or two combs of extra bees are shaken into it to compensate for the flying bees which will leave it. After a day of "flying" weather only young bees will be left in the nucleus and a queen may then be safely introduced to it by any reliable method.

Great care must be taken that the bees cannot pass from one compartment to the other.

Both queens may be left to produce brood until it is desired to unite the two stocks. The old queen is then removed and the bees allowed to mingle slowly by a slight withdrawal of the division board which is later removed altogether.

CHAPTER VIII

THE SNELGROVE "ONE-HOUR" METHOD

To Provide a Fresh Escort for a Newly-arrived Queen.

ALTHOUGH reputable queen-breeders are careful to sell queens only from healthy stocks it sometimes happens that queens are sold from stocks affected by Nosema, Acarine, or Foul Brood diseases of the presence of which the vendor is not aware. It is prudent therefore to examine attendant bees for Nosema and Acarine diseases before introducing a newly arrived queen to a healthy stock. Brood diseases cannot be detected by the examination of adult bees but they may be transmitted by them to healthy stocks in certain circumstances. Even the candy provided for the journey may have been made with infected honey.

If a queen's attendant bees are found to be diseased the queen herself may be infected, especially in the case of Nosema disease, and a prudent beekeeper will not introduce her to a healthy stock.

As a general rule therefore it is desirable to give every newly-arrived queen a clean cage, fresh candy, and a new escort before introducing her by a caging method. Reference has already been made to the practice, advised by some writers, of putting a queen into a cage with newly-emerged bees, and to the obvious disadvantage of this—that such young bees cannot feed the queen; and also to the ease with which a new escort of bees of any age can be provided by means of the "Water" method (p. 131). A still more convenient method is available and has been used by the writer since 1912—many hundreds of times, without a single failure.

It is as follows:—

Confine a few bees in a match-box or empty cage. Keep them dark and without food for from 5 to 10 minutes. Place the queen amongst them and keep them dark for half an hour.

Or in greater detail:—

Place the requisite number of bees—20 or more—in a match-box. This may be most quickly done by placing the box, three-quarters open, over the bees where they are thickest on the comb, and gently closing it with the fingers, the bees being preferably taken from a comb of brood (p. 102). Close the box and put a pin through the side to prevent it from being accidentally opened. Keep the box in the pocket for from 5 to 10 minutes. At the end of this period partly open the box, placing the thumb over the opened part. Drop the queen amongst the bees, close the box, leaving a very narrow slit for ventilation, pin the side, and return it to the pocket for half an hour. Bees and queen may then be put into a cage furnished with food.

Apart from the interval the whole of this operation can be carried out in five minutes or less. This time can be reduced by using a foodless cage instead of the match-box.

Bees confined in the dark and without food become frightened, and are concerned with nothing but a means of escape. They take no notice of a queen introduced to them, and during the succeeding half hour they and the queen may be presumed to attain the same odour, the acquisition of which is promoted by the warmth of the pocket.

Provided the bees are from one stock only, it does not matter whether they are old or young, are taken from a queenless or queen-right stock, from brood or broodless combs, or even if they have fertile workers amongst them. Fertile or virgin queens alike may be introduced to them.

A few minutes after introduction of the queen to the box she may be inspected by pushing the box open under a sheet of glass or celluloid. Bees and queen will be seen to be endeavouring to escape and quite unconcerned with one another.

The safety of this method of introducing a queen to a new escort may be demonstrated by the following interesting experiment:—

Confine some bees in a match-box as previously described. After 10 minutes introduce a fertile queen to them. Five minutes later introduce another fertile queen to the same box. Keep dark for a further 10 minutes or more. Inspect the bees under glass. Both queens will be seen to be quite safe and unconcerned with each other. They and the bees will be running about in the box seeking a way of escape. If kept dark they are safe for an indefinite time, but if exposed to light for a few minutes the younger queen, probably recognising her rival by sight, will attack and sting her.

The match-box method is useful not only for providing a queen with attendants quickly, and at any time, but also for changing her attendants periodically when it is desired to keep her caged for an unusually long time. It also serves as the preliminary part of the method of introduction now to be described.

SIGNS FOLLOWING DEQUEENING:

The removal of the queen from a stock is followed by a period of distress and restlessness easily observable by the beekeeper. For a short time, varying from a few minutes to as much as half an hour the bees go about their duties in the ordinary way, not appearing to have noticed their loss. Even during this period however the nurse-bees have been searching for the queen and soon the sense of her loss is felt by all the bees in the hive. They leave their work, neglect their young, and run anxiously over the combs, enquiring of one another,

and exploring every part of the hive. Some emerge
from the hive entrance and run in all directions over
the floorboard and the front of the hive. Unsuccessful
in their search they return quickly to the hive, only to
be followed by other searchers in increasing numbers.
The commotion, which is usually pronounced at the
end of one hour, increases, and continues for about
24 hours, after which the bees, having begun the
construction of queen-cells, become less agitated and
gradually resign themselves to their loss.

During the summer, when the bees are much in movement, they realise the loss of their queen very quickly.
In cold weather they take longer, but it is only in rare
cases that their distress is not apparent at the front of
the hive within an hour. Disturbance by smoking or
shaking during examination for dequeening causes them
to discover their loss more quickly. Their distress abates
at once if their own queen be restored to them within a
few hours, but a strange queen will be accepted only if
they are tricked into believing her to be their own.

After a few hours of orphanage the bees begin to
raise queen-cells and thenceforth look forward to the
emergence of young queens. During this period of
expectancy they are favourable to the reception of a
queen suitably introduced, especially if they are first
deprived of their queen-cells.

The most favourable times, therefore, for the introduction of a new queen are:—

(1) When the bees are in a state of acute distress
    following the loss of their queen.
(2) About 24 hours after their loss, when queen-cells
    have been started.
(3) When queen-cells are fairly well advanced and the
    bees are expecting a new queen.

(1) is favourable for direct introduction; (2 and 3) for
introduction by any method.

SNELGROVE "ONE-HOUR" METHOD.

We have now to consider a method of introduction which is easy and expeditious, reliable in all circumstances and which involves only one visit to the hive.

This was first used by the writer in May 1935. A small stock had been left queenless for 3 days. A few of the bees were placed in a match-box and kept in the pocket for 10 minutes. An alien queen was placed with them, and after being kept in the pocket for half an hour was allowed to run down into the hive by way of the feed-hole. She was subsequently found to be laying well.

This experiment was repeated several times during the year, the queens being introduced to the boxes of bees after 5 minutes confinement and run into the hives at the end of half an hour. It was hoped that the method might be called the "Half-hour" method. The experiments were carried out at various times of day and whilst the majority were successful there were some failures. It was observed that after a half-hour interval after dequeening some stocks showed little or no sign of the distress which appeared to be an essential condition of success.

In the following year 59 experiments were made, the period of queenlessness being lengthened to one hour. The introductions were still carried out at all times of the day, even during the robbing period, and in several cases the hives were examined on the same or the following day to ascertain if introduction had been successful. Notwithstanding these imprudences, of the 59 experiments made 53 were successful—that is to say the queens were subsequently seen parading quietly amongst the bees on the combs. Four of these afterwards disappeared—doubtless on account of premature examination of the hives.

Careful consideration of the failures suggested that they were probably attributable to:—

(1) Introduction during the day time when incoming field bees would not have experienced the state of queenlessness.
(2) Robbing.
(3) Premature examination.

In the following year 29 experiments were made under the following conditions:—

(1) Queens were introduced to the boxes of bees after the latter had been confined for 10 minutes.
(2) Introduction to the hives was effected about 50 to 60 minutes later.
(3) The operations were carried out late in the evening.

All these experiments were successful, the queens being introduced in varying circumstances, including the following:—

(1) To stocks long queenless (up to 8 weeks).
(2) To stocks which had been headed by drone breeding queens.
(3) To stocks containing fertile workers.
(4) When queens had been caged up to 12 days.
(5) To one or both sides of "duo" hives.

In some cases of (1) (2) and (3) worker brood was given to the stocks a day or two previously.

In three cases of requeening one side only of "duo" hives the queens showed by torn wings that they had been encased and subsequently liberated. This did not happen when both sides were requeened simultaneously. The requeening of "duo" hives is recognised as difficult. By the writer's Water method however (p. 130) it is easy to requeen either one side, or both sides at one time.

Directions for introducing by the "One-Hour" method are as follow:—

At a convenient time before dusk—

(1) Dequeen the stock.

(2) At the same time take from 20 to 30 bees from a comb (preferably, but not necessarily, from one containing brood), and put them into a safety match-box (p. 135). Keep this in the pocket for about 10 minutes and then place the new queen, taken directly from combs or from a cage, in the box with the bees. Return the box to the pocket.

(3) About 50 minutes or an hour later open the match-box under the quilts, or over the feed-hole, so as to allow queen and bees to run down on to the combs. A preliminary puff of smoke, such as might be given from a cigarette, should be given to the hive bees before the box is placed in position, to clear the way for the queen to run down. Leave the box in position and cover it at once with the quilts, or, if it is over the feed-hole, with a piece of dark cloth or the deep lid of a tin. Avoid unnecessary disturbance and noise.

(4) Do not examine the hive for a week.

The times given above need not be strictly adhered to. The queen may be put into the match-box at any time after the bees have been confined for 5 minutes, and the 50 minutes may be exceeded by any length of time during the same evening, although it is not desirable to keep the queen without food for much more than an hour.

The success of this method depends on four conditions:—

(1) The queen acquires the odour of the hive bees before she is introduced.
(2) She is hungry when liberated.
(3) She enters the hive with a friendly escort.
(4) The hive bees are in a state of distress and are anxiously looking for their queen.

As before stated a stock usually shows great distress and commotion within an hour of losing its queen. In some cases, however, especially in large stocks with two brood chambers or where the bees have begun to cluster in autumn, it is as well to disturb the bees when removing the old queen in order to cause them to realise their loss more quickly (p. 137). If distress is not pronounced at the end of one hour, which is rare, the bees should be disturbed and the introduction delayed for a further half hour.

Within a minute or two after the introduction of the queen the commotion at the hive entrance ceases. The bees which were running over the alighting board and up and down the front of the hive suddenly cease their search and joyfully enter the hive where a welcoming hum soon develops into a pleasant roar which tells the beekeeper that the introduction has been successfully accomplished.

Apart from the intervals the whole process takes but a few minutes and the new queen, if laying, resumes her maternal functions during the same night. No equipment, except the homely match-box, is needed, and there is practically no disturbance of the work of the bees.

For precautions to be taken in the case of travelled queens see p. 40.

A number of hives may be requeened by this method during one visit to an apiary, in which case the time taken for each hive is greatly reduced. The beekeeper however must be methodical and careful not to mix boxes containing deposed and young queens in his pockets. The boxes should be labelled and those containing the old queens put safely away. The boxes containing the young ones may be kept between the warm quilts of their respective hives, instead of in the pockets, during the intervals. To avoid mistakes the matchboxes should be labelled beforehand and the description of the queens and the numbers of the hives for which they are intended carefully recorded.

SNELGROVE RECTANGULAR MULTIPLE CAGE.

Reference has already been made to the possibility of introducing two queens to one lot of bees and to the fact that they are safe whilst kept in darkness (p. 136). It was a natural inference that if the queens were kept apart it would be possible to introduce an unlimited number of queens and to keep them safely for an unlimited time whether kept in darkness or exposed to light. Experiments which the writer carried out in 1922 were successful and in that year he showed Mr. Cowan, at Clevedon, three queens in one cage, all quiet and contented, and being well cared for by about three dozen bees. A letter describing this phenomenon was sent to the Editor of the " British Bee Journal " in November 1923. It was kept in the office safe until October 1936 when it was opened and read at the annual conversazione of the British Beekeepers' Association. In his lectures on Queen Introduction the writer has several times exhibited cages containing from three to six queens, the queens in each case being fed by one lot of bees. These cages always prove to be fascinating to an audience.

The queens are placed in separate compartments to which the bees have access through excluder zinc. If the compartments were merely separated by excluders, so that the queens could reach one another with their antennæ, they would experience mutual hostility and the bees would soon kill them all except one. For this reason it is usually impossible to winter two or more queens in a hive if they are merely kept apart by means of excluders. During the summer months however two queens separated by an excluder may be tolerated in the same hive for a long time.

The method of separating the queens in the writer's multiple cage may be best explained by reference to the illustration on page 145. This cage is made to house three queens only but the same principle of construction would apply to cages designed to hold any number of queens.

The body of the cage is a light box, made of three-ply wood, of external dimensions $5\frac{3}{8}''$ x $2\frac{1}{4}''$ x $1''$. This is divided into four compartments by thin wooden partitions. Three of them (Q) are for the queens, and the narrowest, $\frac{1}{2}''$ wide, contains a tin trough (CTR) for candy. This trough is shorter than the width of the cage by $\frac{1}{2}''$.

A piece of excluder zinc (Ex) carefully selected and cut, is let into the partitions at a distance of $\frac{1}{2}''$ from the front of the cage to form a corridor (CR) outside the queen compartments. Openings (O) are cut through the partitions in the corridor to allow the bees to pass from one end of the cage to the other. A hinged piece of sheet metal (H) closes or opens the passage to the candy. Ventilation holes (VH) covered with perforated zinc or wire cloth, are cut in the back wall of the cage. Each compartment is opened or closed by means of a sliding lid of celluloid (L) which permits of observation or removal of the queens and bees, or shows when the candy supply needs renewal.

The construction of the cage can be simplified, the essentials being separate queen compartments into which the bees pass from the corridor through queen excluder, and a compartment for food. Another set of queen compartments may be provided on the opposite side of the corridor.

The greater the number of queens and bees to be kept in the cage the larger must be the supply of candy, and a candy compartment at each end conduces to a more even distribution of the bees to the compartments. The corridor should be relatively small so that the greater number of the bees must at any time be in the queen compartments.

The apparent contentment of bees and queens in these cages is very striking. They can be exposed to light and observation for an indefinite time and make little or no effort to escape. When not under observation however it is desirable to keep them dark as in the case of all other queen cages, and as far as possible to keep them in an equable temperature such as that of an ordinary living room. They must have ample ventilation at all times. This may be increased in warm weather by slightly withdrawing the lids.

To stock the cage proceed as follows:—
Remove the wire cloth coverings of the ventilation holes and temporarily substitute corks for them. These should be easily movable. Shut off or remove the candy. Withdraw the lids of the queen compartments almost completely. By means of a postcard scrape sufficient bees (about 30 per queen) from a brood-comb into the open cage and close the lids immediately. Keep the cage dark in the pocket for a quarter of an hour or more. Still keeping the cage dark by means of a piece of cloth, introduce a queen through each ventilation hole at the back of the cage, replacing the cork immediately after doing so. Replace the cage in the pocket for about half an hour and then give the bees access to the candy. Remove the corks and restore the wire cloth to the ventilation holes.

If preferred the queens may be introduced by the water method (p. 131). For this it is not necessary first

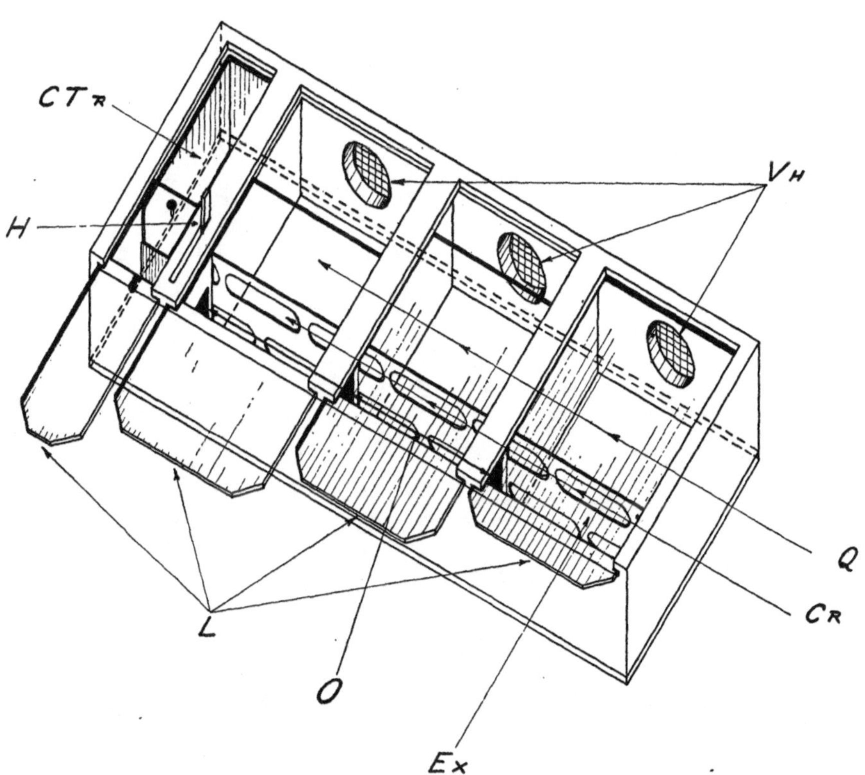

SNELGROVE RECTANGULAR MULTIPLE CAGE.

K

to confine the bees for a quarter of an hour, but it is preferable to do so and they should be kept dark and without food for the subsequent half hour.

Surplus queens may be kept for weeks in a cage of this kind. Any one of them may be easily removed for use at any time, and, if desired, another may be inserted in her place by the water method.

As previously stated it is not desirable to keep queens in confinement longer than is absolutely necessary. Their nurse attendants become too old for their duties and later die of old age. A week should usually be regarded as a maximum time in ordinary circumstances, but if it is desired to confine them for a longer time they should be removed after about 10 days and the cage stocked with fresh bees before they are re-introduced to it. Care should be taken to replenish the candy store at intervals. It will be found that this diminishes rather quickly on account of the number of bees in the cage.

A multiple cage is not only of great interest to an audience and ideal for the preservation of surplus queens but it is also suitable for transmission through the post. On October 24th last, a cage containing three queens and about seventy bees was posted by the writer to Mr. A. G. Pilkington at Milborne Port. After inspection he re-posted it on the 25th and it was received back on the 26th. The queens were lively and well, although the weather was cold, but the bees had consumed all their candy and were hungry. For travelling therefore a larger candy trough would be necessary.

SNELGROVE CIRCULAR MULTIPLE CAGE.

A more elaborate form of this cage, made for the writer by Mr. G. H. Logan, is illustrated in Pl. X. Essentially it consists of a circular box of 6″ diameter and $\frac{3}{4}$″ depth, divided into six sectors. The main portions of the sectors (S) are the queen compartments. The small inner portions (C) are enclosed by a circular strip of queen excluder (EX_1) of diameter $1\frac{1}{2}''$, the enclosed space corresponding to the corridor in the rectangular cage. This space is small so that the majority of the bees congregate in the queen compartments. The compartments are enclosed by a circular piece of excluder (EX_2) so as to provide for a second corridor inside the wall of the cage.

A vertical spindle (SP) is passed through the centre of the base into a rotating wooden collar (RC) underneath the base, and openings (OP) are made through the base and collar to allow the bees to descend to a glass jar (J) filled with candy, which fits tightly into the collar. Rotation of the collar closes the openings and prevents the bees from escaping whilst the candy-jar is being changed. The whole is mounted on a stand.

The top of the cage is covered by a circular sheet of stout celluloid through which the queens may be observed. This cover rotates about the central spindle at a height of $\frac{3}{4}''$ above the floor of the cage. One sector (SB) is cut away from it and the space so made is surmounted by another rather larger sector (SV) which also rotates on the central spindle. When this sector is over the space in the cover below, the cage is closed. When it is moved to either side one of the queen compartments is open and the queen can be removed if this is desired. Any one of the queen compartments can be opened by rotating the lower cover until the cut-away space is over it and then moving the upper sector away from the same space. Ventilation holes (VH_1) of $\frac{3}{8}''$ diameter, one for each queen compartment, are cut in the wall of the cage and these are covered by perforated zinc. Additional small ventilation holes (VH_2) are pierced in the base of the cage. When not under observation the wooden lid (L) keeps the bees in comparative darkness. Complete darkness can be secured by covering the whole cage with a piece of dark cloth. The candy

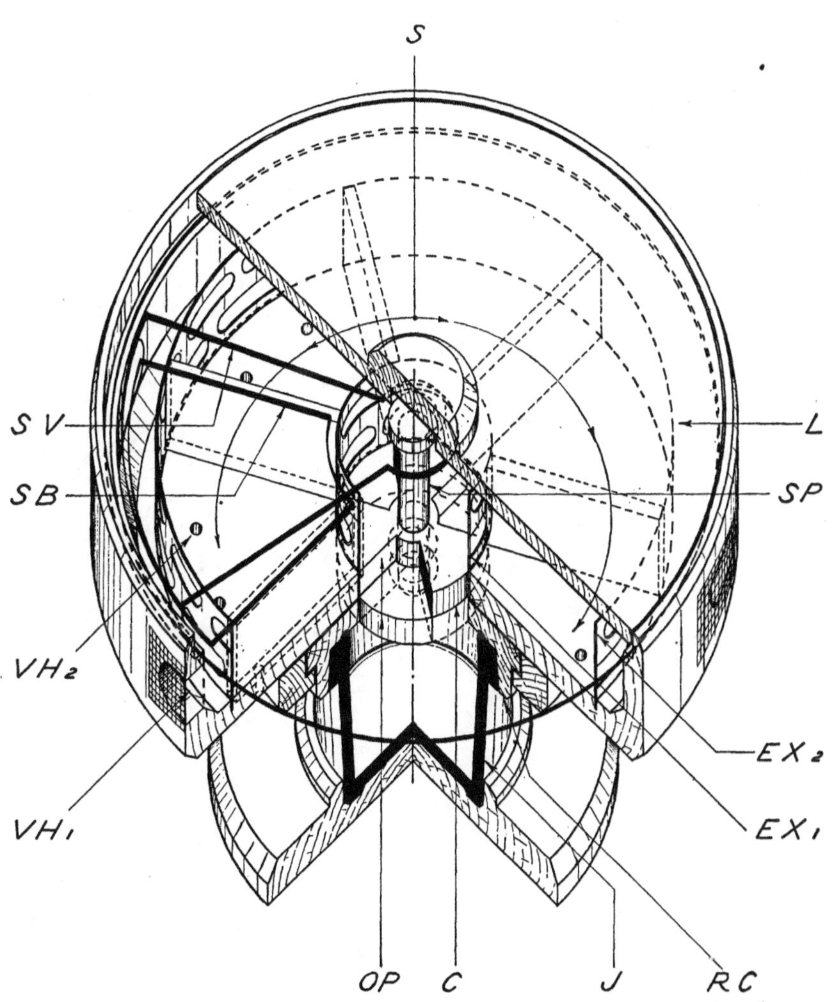

SNELGROVE CIRCULAR MULTIPLE CAGE.
(PARTS CUT AWAY TO SHOW CONSTRUCTION.)

supply is visible at all times and is renewed by substituting a full jar for an empty one.

In the Autumn of 1936 the writer kept six marked queens, five old and one young, in this cage from early October until Christmas. The cage was kept on a shelf in a room which was centrally heated so that the bees did not experience cold. The temperature of the room varied with the weather however and on mild days the bees were restless, which was not good for them. They were restful when the temperature of the room was round about 60° F. Occasionally, on mild days, they were taken out of doors, well covered, and allowed to take cleansing flights from one of the ventilation holes. Some of the bees, but not all of them, would fly and return in the usual manner. As they were mostly young when put into the cage comparatively few of them died during the three months, and the cage was scarcely soiled. Without the cleansing flights it would have been necessary to change the bees several times during the period. The bees were wearing out in December and it was intended to change them at a favourable opportunity and to attempt to winter the queens. To the writer's regret however the cage was put out for a cleansing flight on Christmas Eve and forgotten for some hours. Both bees and queens were hopelessly chilled and the experiment was discontinued.

Some interesting observations were made in the use of these multiple cages:—

(1) Pollen, freshly taken from a comb and placed in a cage was slowly consumed.

(2) Water was sometimes given on a small sponge but the bees were never seen to take it. It may be that bees need water, as such, only when they have brood to rear.

(3) The bees passed freely from one queen to another.

(4) At first the bees were fairly well distributed

amongst the queen compartments but in course of time they showed a marked preference for the young queen. There were many more in her compartment than in those of the oldest queens and occasionally one of the latter would be left with a single faithful attendant, but never without one. Disturbance would cause the bees to redistribute themselves. If no young queen were present in the cage the preference was much less marked but some of the queens would usually be attended by more bees than the others, and the favoured one of one day might be the comparatively neglected one of another. In the rectangular cage the queen furthest from the candy was often the least well attended. This suggests that a candy trough at each end of the cage would be an improvement.

These cages were invented by the writer mainly as objects of interest but the fact that they show that an unlimited number of queens can be introduced to one lot of bees, and that the queens are well fed and live contentedly for a long time in such circumstances, suggests the possibility of wintering several queens in one hive. This has not proved to be practicable when they are merely introduced to combs encased by excluder zinc and separated by intervening combs. The bees ultimately select one and destroy the others, and even if they did not do so the seasonable movement of the cluster would prove to be a difficulty. The surviving queen appears not to suffer from her winter confinement and lays normally in the spring and during the ensuing season.

By utilising the principle of the multiple cage the writer has succeeded in wintering three queens in a very large stock, the queens being kept far apart but the bees being able to go from one queen to another.

In this case however the bees resolved themselves into three separate colonies during the winter, re-uniting without loss of queens in the late spring. Experiments are being made with the object of wintering several queens, if possible, in the one brood nest of an ordinary stock.

During the early months of the year young queens are often needed to repair winter losses but in countries like our own they are almost unobtainable. They can sometimes be wintered in well provisioned nuclei but this involves much work and the provision of additional equipment. Any method of wintering several queens in one hive without serious detriment to them would prove to be a great boon.

It is commonly believed that queens are harmed by prolonged confinement in cages. This has not been confirmed by the writer's experience of a few cases. Observed deterioration in a queen's after life may be due to rough treatment of a cage, e.g. in the post, or to malnutrition whilst she is caged.

Doolittle[1] who subjected some caged queens to rough treatment by throwing them about his shop concluded that the subsequent deterioration of some of them was due, not to the rough treatment, but to the sudden and enforced cessation of their egg-laying.

Miss Betts cites Dr. Brünnich[2] as saying, "It is a fairy-tale that a queen cannot be kept in confinement and caused to stop laying without being injured." She states also that "Allgayer has observed this too, having accidentally confined mated queens for 18 and even 24 days without their suffering any perceptible harm." She further refers to B. D. Milojevic[3] of Belgrade, who "kept a young mated queen for four months in a cage with a few workers, at room temperature. At the end of that time the queen could lay

---

[1] Doolittle, G. M., *Queen Rearing*, p. 103.
[2] Betts, A. D., *The Bee World*, Sept., 1938, p. 99.
[3] Betts, A. D., *The Bee World*, April, 1938, p. 43.

quite normally and was kept under observation further during April—August in a strong colony."

Raynor[1] in 1880 exhibited a queen which had been in confinement for seventeen weeks, her escort of 200 bees having been reduced to 50 by deaths from old age. He described her as being as sprightly and active as on the first day of imprisonment.

It should not be inferred from the description of the caging experiments in this chapter that the writer advocates or approves the prolonged caging of queen bees when this can be avoided. As a general rule it should be considered that a queen should not be confined a day longer than is absolutely necessary.

[1] Raynor, G., *Queen Introduction. The Ligurian Bee.* A paper read to the B.B.K. Association, January, 1880. p. 10.

CHAPTER IX

OTHER METHODS

IN addition to the methods of queen introduction already described in this book there are others which have from time to time been recommended but which are little known and used. Some of them are merely modifications of well-known methods whilst others are worthy of notice as suitable for experiment and improvement. Brief descriptions of some of these will now be given.

1. TAPPING OR BEATING HIVE.

If a queenless hive be closed and tapped or beaten vigorously for some time, as in the manner of driving a skep, the bees will be reduced to a state of terror and will accept a queen placed amongst them. This is essentially similar to the "Smoke" Method, but, as Maisonneuve observes, it involves long and fatiguing work and therefore cannot be regarded as a satisfactory method of introduction.

2. SADLER'S METHOD.

Cheshire[1] referring to Vol. XIV of the " British Bee Journal," p. 332, says, "Mr. Sadler asserts that if, after the abstraction of the old queen, thin sugar syrup be scented with oil of peppermint and poured between the combs from a vessel with a small spout, a stranger then put on the top of the bars and wetted with the same syrup will be at once accepted."

This method is doubtless effective, but the Water Method is much simpler.

[1] Cheshire, F. R., *Bees and Beekeeping*, Vol. II., p. 344.

3. COBB METHOD.

Mrs. A. A. Cobb[1] finds the following method "better than any other" she has tried. The hive is dequeened and any queen-cells that may subsequently be started are removed. The new queen is confined in a perfectly clean and empty cage and this is placed, screened side down, over and between two brood combs for 36 hours. The screen over one of the end holes of the cage is then removed with the least possible disturbance so as to liberate the queen.

Apparent disadvantages of this plan are:—

(1) A 36-hour interval is rather inconvenient for the beekeeper.
(2) The queen may be without food for a considerable time.

4. SCENTED SYRUP METHOD.

Meurant[2] gives the following method, which he describes as almost infallible. It is a variation of the Honey method described on page 126:—

When the bees have ceased flying for the day place the queen in a small flat cage which can be easily opened. Prepare a scented syrup of honey or sugar, warm it to about 30° C. (86° F.), and plunge the cage containing the queen into it. Sprinkle the bees of the hive with the syrup, place the cage between two combs, open the cage and close the hive. The addition of scent to reduce bees and queen to the same odour is doubtless an advantage, and the caging of the queen whilst immersing her would be preferred by beekeepers who are not skilled in handling queens.

In describing the use of a cage of his own design embodying the Asprea principle (p. 73), Meurant recommends that the cage and hive bees be perfumed

---

[1] *Gleanings in Bee Culture*, 1937, p. 615.
[2] *La Gazette Apicole*, 1935, p. 253.

for two or three hours before the queen is put into the cage. After the queen has been exposed to the perfume for about 4 hours the bees are admitted to the cage through excluder zinc and their attitude is observed. If it is favourable the queen is liberated. In this way Meurant considerably reduced the period of the queen's confinement as compared with that necessary in the Chantry, Pieyre, and similar cages.

He recommends oil of balm as most suitable for a perfume, saying that its odour resembles that of queens and is pleasing to the bees.

5. REQUEENING BY MEANS OF AN ARTIFICIAL QUEEN CELL.

Otto Schulz[1] devised a method of introducing a queen by means of a large artificial queen-cell. The latter was made in the usual way by means of a dipping stick, the effective portion of which was about $1\frac{1}{4}''$ in length and $\frac{3}{8}''$ in diameter. The end of the cell was thinned and closed, and was pierced with pin-holes so that the imprisoned queen might have air and the bees easily nibble through it. The queen was introduced to the cell through its open base and the cell then attached to a wooden cell holder.

The cell was suspended between brood combs of a queenless stock. Whilst releasing the queen the bees would feel that they were assisting in the "hatching" of a young queen. When released the queen would be hungry and therefore in a state favourable to good reception. Maisonneuve reports several successes with this plan but there is always the danger that the queen might not be liberated until she is exhausted by hunger. In any case the amount of work is considerable and unnecessary in the light of easier methods.

[1] Perret-Maisonneuve, A., *L'Apiculture Intensive et l'Elevage des Reines*, p. 294.

6. Introduction to Colony Containing a Drone-Laying Queen or Fertile Workers.

Keen[1] recommends the following plan:—

Prepare a hive containing a comb of brood in all stages and some uncapped honey—but no bees. Using a press-in cage place a fertile queen over some uncapped honey or brood.

Take the drone breeding stock to a distant part of the yard and put the prepared hive in its place. Shake the drone-breeding bees from their combs. They will return to their old location and will soon be endeavouring to release the new queen in the prepared hive.

This method would be satisfactory where a drone-laying queen is to be discarded but would be uncertain in the case of fertile workers, some of which would be likely to return to their old home (p. 58).

7. Requeening without Dequeening.

Réaumur[2] records some successes in introducing queens to stocks headed by their own queens. He merely placed the new queen amongst the bees and observed not only that they were well received but also that they began to lay. His experience was unusual for queens so introduced are almost invariably balled and killed.

Miller[3] succeeded in introducing new queens into stocks in which the reigning queens were laying by means of the "Smoke" Method. He stipulated that conditions should be favourable and especially that the reigning queen should be engaged in laying in one part only of the hive. The new queen was introduced to the other part so that both queens might continue to lay until they met. In some cases, he says, this might not happen until the bees clustered for winter.

[1] Keen, G. H., *Introducing Queen Bees*, p. 13.
[2] Réaumur, K. A. F. *Mémoires pour servir a l'Histoire des Insectes.* Vol. V, p. 639.
[3] Miller, Dr. C. C., *Fifty Years Among the Bees*, p. 296.

If favourably received and tolerated until after she had begun to lay the young queen would ultimately kill the old one.

Such experiments as these, however, are most unsafe and should be avoided where valuable queens are concerned.

If it is desired to substitute a young queen for an old one without first removing the latter it is best to divide the stock into two portions by means of a close fitting division board or a Snelgrove Swarm Control board (p. 132), and to introduce the young queen to the queenless part. When the young queen has established a fair brood nest the two portions of the stock may be united. The young queen will kill the old one when they meet.

G. S. Demuth's[1] method of requeening without dequeening was to give a ripe queen-cell to a food chamber (stored super) set off on a separate stand. This was done at the time when the honey supers took the place of the food chamber and when the latter contained a little brood. After the close of the honey season and removal of the honey the food chamber, now containing a young fertile queen, was replaced on the parent stock with a queen-excluder and sheet of newspaper between the two stocks. In course of time the bees, in almost every case, would destroy the old queen, and the young one would reign in her stead. The excluder was then removed.

The assumption that the young queen always kills the older one is sometimes challenged. There may be cases where the older queen survives but in a limited number of experiments carried out by the writer with marked queens in 1938 the young queen survived in every case.

[1] Root, A. I. and E. R., *ABC and XYZ of Bee Culture*, p. 621.

8. Paper Bag Method.

G. H. Keen[1] of New Zealand describes a method which can be used "with the minimum of risk". A thin paper bag, such as is used for sweets, is pierced in several places with a fine nail, and the queen is put into it without an escort. The bag is then opened under a cluster of bees belonging to the hive into which the queen is to go, and by suitable movement of the edge of the bag about a "tablespoonful" or more of the bees are caused to drop into it. The bag is then closed, care being taken that the queen does not escape—and suspended between two combs of brood "where most bees are".

Mr. Keen states that the bees in the bag will "chew their way out" and that "the hive will continue in the even tenor of its way", which means, presumably, that the queen will be successfully introduced.

It is much more likely, of course, that the external bees chew their way into the bag and so liberate the queen.

It would appear that if they do this too quickly, before the queen acquires the odour of the bees, or before the latter have realised their queenlessness, the queen will be in danger of being balled. On the other hand she may be imprisoned for a long time, without food, before she is liberated. On one occasion the writer experienced a case in which bees refused for two days to liberate a queen by nibbling away thin paper. They built queen-cells instead.

Apart from the special care needed in choosing and manipulating the paper bag, the defect of the method, which is otherwise commendable, is the uncertainty as to the time when the queen will be liberated.

[1] Keen, G. H., *Introducing Queen Bees*, p. 11.

9. INTRODUCTION TO AN ARTIFICIAL SWARM.

Freudenstein[1] recommends the following method as suitable for a precious queen:—

Take a well-ventilated swarm box and place on its floor a cage containing the queen to be introduced. The cage is to be provided with candy by means of which the bees can liberate the queen. Shake into the swarm box about 2 Kg. ($4\frac{1}{2}$ lbs.) of bees from brood combs of one or more stocks. Keep the swarm dark and cool in a cellar for two days and then hive it as a natural swarm on starters or full combs in the usual way. Feeding will be necessary unless a honey flow is in progress.

10. INTRODUCTION TO AN ODORISED ARTIFICIAL SWARM.

A method, similar in essentials to the one described above, is given by a writer, presumably the Editor, in the "Deutscher Imkerführer" of December, 1937. It is as follows:—

At a time when bees are flying suspend combs of bees, taken from one or more stocks, but without a queen, on a comb-rack in the open air. Leave them thus for ten minutes. (During this interval they will tend to lose their hive odours and some of the older bees will return to their hives.)

The queen to be introduced is placed alone in a cage fitted with candy and the cage is fixed to the inner side of the lid of a ventilated swarm box.

The exposed bees are first sprinkled lightly with water from a small watering-can, the water having been previously odorized by the addition of thymol—10 to 12 drops to a pint.

About 3 lbs. of bees are then shaken from the combs through a funnel into the swarm box which stands on

[1] Freudenstein, Dr. K., *Wege zur Völker Vermehrung, Deutscher Imkerführer,* Juli, 1937, p. 100.

L

a weighing machine. The box is closed and kept in a cool dark place. The bees release the queen in due course and after 24 hours the artificial swarm is hived as a natural one and treated as such afterwards.

Other interesting methods given by the same writer are as follow:—

11. Introduction to an Odorised Nucleus.

Take from one or more stocks a few (three to five) combs of emerging brood, with bees, but without a queen, and after sprinkling them with thymol water put them into an empty hive. Add two other combs containing honey and pollen. Move the stock away and, if necessary, add to it more young bees, moistened with thymol water. After 10 minutes insert the caged queen, the cage being provided with exposed candy. Pack the stock warmly and feed in the evenings after the second day until there are sufficient flying bees.

12. "A Quick Method."

The stock is de-queened and the combs suspended on a rack in the open air, two inches apart. The new queen is placed alone in a cage provided with exposed candy, and the cage smeared with honey from one of the combs. When the bees show distress at the loss of their queen fasten the cage to the middle of one of the brood combs. Wait until a bunch of bees surrounds the cage—meanwhile sprinkling returning field bees with water—and then replace the whole stock in the hive.

It is hardly necessary to remind the reader that methods involving the prolonged exposure of honey combs in the open air should be employed only when there is no danger of robbing. This method can hardly be called a quick one. Like the preceding ones it involves much work and is not to be recommended.

13. Presentation of a Shaken Artificial Swarm to a Queen.

Place the queen in a cage provided with exposed candy. Fasten the cage to the inside of the lid of a ventilated swarm box. Place the lid in a convenient position (e.g. suspended and tilted near the ground, cage underneath) and shake the queenless bees from their combs quite near to it. (This should be done on the site which the bees have been accustomed to occupy.) The bees will soon cluster as a swarm round the queen in her cage. The swarm is kept in its box in a cool dark place until the evening by which time the queen will be liberated. The bees may then be hived as a natural swarm in the usual way.

14. Manley's Method.

Manley[1] gives a method of substituting a fertile queen for a virgin queen which has failed to mate and in whose stock no unsealed brood remains. He removes the virgin queen and waits until the bees show distress at her loss—usually from one to three hours. The attitude of the bees is first observed by presenting the caged fertile queen at the entrance. If this is favourable she is placed on the alighting board and allowed to run into the hive where she will almost always be accepted. Mr. Manley adds that if a virgin be removed in the afternoon of one day the stock will be ready to take any queen it can get next morning. When they "scent her they will literally pour out of the hive and surround the cage." She may then be released (p. 115).

15. "Self-Requeening."

Demuth and Root[2] observed that when a queen was allowed the full range of two full-depth Langstroth

---

[1] Manley, R. O. B., *Honey Production in the British Isles*, p. 274.
[2] *Gleanings in Bee Culture*, 1936, p. 205.

brood-chambers throughout the spring, summer, and fall, the stock usually requeened itself near the close of the honey season. Supersedure at this time would not result in loss of honey. A smaller hive would not induce supersedure at the best time; if this happened in the following spring it would result in a considerable weakening of the stock prior to the honey flow.

As previously stated a very large hive exhausts the fertility of a queen sooner than a small one and is therefore conducive to early supersedure.

There are some British beekeepers who rely on natural supersedure but unfortunately the British climate does not often permit of the use throughout the year of a brood chamber of the capacity described by Root.

CHAPTER X

INTRODUCTION OF VIRGIN QUEENS

IT is well known that it is more difficult to introduce a virgin than a fertile queen.

Virgin queens are plentiful during the swarming season, and can easily be bred in considerable numbers throughout the summer. They are often advertised at low prices and are usually introduced to nuclei for the purpose of being mated. Sometimes they are introduced to full stocks for the purpose of requeening notwithstanding the risks that they may not be accepted or may fail to mate. As a rule, full stocks should be requeened with fertile queens.

Many of the methods given for the introduction of fertile queens do not apply to virgin queens. It is therefore desirable to consider the circumstances in which the latter are accepted by queenless bees.

1. CONDITIONS OF THE STOCK.

A stock may best be induced to accept a virgin queen *when it is expecting one*, that is to say when it has built queen-cells after being made queenless. The more advanced the queen-cells the more natural will it be for a virgin queen to appear in the hive. The ideal time for introduction therefore is just before the time when the young queens may be expected to emerge from their cells. No eggs or very young larvae—both deterrent to successful introduction of virgin queens— will then be present.

In nature—e.g. after natural swarming—the presence of a young virgin queen is consistent with the presence of sealed brood only, and it is therefore desirable, in cases where bees have been queenless for some time, and are consequently broodless, to provide them with some sealed brood before the introduction of a virgin queen.

The least favourable time for the introduction of a virgin queen is immediately after the removal of a fertile queen, for then the presence of eggs gives the bees an opportunity—which they usually take—of rearing their own queen. Occasionally, however, a virgin queen, returning from a successful mating flight and entering the wrong hive, is accepted by the bees notwithstanding the presence of young brood and a reigning queen. The latter is deposed and the intruder reigns in her stead.

A stock deprived of its queen and all its brood is soon afterwards in a favourable condition to receive a virgin queen (p. 116).

A strong stock is more likely to reject a virgin queen than a weak one. Hence it is usually easy to introduce a virgin to a nucleus stock, and even more easy to introduce one to a baby nucleus.

Ripe queen-cells should be broken down before a virgin is introduced. If queen-cells are unsealed it is desirable, although not always necessary, to destroy them also.

Bees taken from a stock from which a swarm has issued will readily accept a virgin queen.

A preliminary period of queenlessness, varying from 6 hours in the case of the "Shaken bees" method to as much as two weeks for other methods, is generally favourable to successful introduction.

2. CONDITIONS OF THE QUEEN.

(a) Age.

It is stated[1] that Langstroth was the first to ascertain that the best time to introduce a virgin queen is "just

[1] Langstroth, L. L., *The Hive and Honey Bee*, p. 286.

after her birth,—as soon as she can crawl readily". At that age she presumably has no special odour or other regal characteristic. She may be liberated on a comb amongst the bees or allowed to crawl in at the hive entrance. Herrod-Hempsall[1] uses the expression "a few hours old" in this connection. Sladen[2] gives "one and a half to three days" as the best age for introducing a virgin queen to a nucleus specially formed to receive her. Root[3] after stating that a young virgin, just emerged, generally weak, can usually be let loose in a queenless colony without caging and be favourably received, goes on to say that "one from two to six days old is as a rule much more difficult to introduce than a laying queen" and that "one ten days old, more than old enough to be fertilised, is most difficult."

Dr. Miller[4] introduced virgin queens, less than one day old, to stocks without first removing the old queens. He reported good success with some stocks but failures with others.

These somewhat divergent views all lead to one conclusion—that the younger a virgin queen the more likely is she to be well received. Introduction by caging her in the hive for several days, which is usually recommended, increases her age and, in this respect only, makes her less acceptable to the bees. For this reason, if for no other, direct methods of introduction are to be preferred.

(*b*) Activity.

Virgin queens, except when newly emerged, are much more active than fertile ones. They are more easily frightened by interference and when liberated amongst strange bees are apt to run quickly over the combs. Not only do they thus invite attack by the

---

[1] Herrod-Hempsall, W., *Beekeeping, New and Old*, Vol. I., p. 700
[2] Sladen, F. W., *Queen Rearing in England*, 1913, p. 27.
[3] Root, A. I. and E. R., *ABC and XYZ of Bee Culture*, 1929, p. 473.
[4] *L'Apiculture Française*, 1937, p. 255.

bees, but occasionally they will run out at the hive entrance and perish. For this reason it is sometimes advisable to introduce a virgin queen towards nightfall, when there is little activity amongst the bees, and to close the hive entrance until next morning.[1] Experience seems to show that confined bees are less likely to attack a newly introduced queen—whether virgin or fertile—than are those which have liberty to fly. They are probably too much concerned about their imprisonment to pay undue attention to a queen.

METHODS OF INTRODUCTION

1. CAGING METHOD.

The usual caging methods (Chap. V) may be used for the introduction of virgin queens but with two important modifications:—

(1) The queen must have access to pollen as well as candy or honey whilst she is confined in the cage because she feeds herself and is not fed by the bees of her escort.

(2) The time of exposure of the imprisoned queen to the bees must be increased.

In the case of "Press into comb" cages it is usually possible to give the queen access to ripe brood, honey, and pollen. When stored cages are used a little pollen should be mixed with the candy.

The cage should be exposed to the queenless bees for at least three days before she is liberated. At the end of this period the bees will have made considerable progress in the construction of queen-cells and will therefore be anticipating the appearance of a virgin queen. If there is no young brood no queen-cells will be made and the young queen will be the more readily accepted because the bees cannot rear one of their own.

[1] Perret-Maisonneuve, A., *L'Apiculture Intensive et l'Elevage des Reines*, p. 271.

Whilst it is usual to recommend the caging of a virgin queen in a queenless hive for three days, Gillet-Croix[1] advises four days. The latter period is the more favourable to success, but as Gillet-Croix rightly observes, the caging method is not always successful in strong colonies and sometimes not even in weak ones.

Maisonneuve[2] recommends that a stock be left queenless for a whole night before the introduction of a caged virgin queen and that two days later the bees be given access to the candy so as to liberate her by the third day.

If queen-cells are broken down an hour or so before a queen is due to be released the chances of successful introduction are increased.

2. Successive Introduction by Caging.

Queen breeders often have more virgin queens than nucleus hives in which to house them. The extra virgins may be inserted into nuclei already provided with queens. The procedure is as follows:—

A virgin queen, usually in the nursery cage in which she has "hatched", and in which is a store of candy and pollen, or in an ordinary introducing cage, is placed on the bottom bar of one of the frames. After three days or more, the fertile queen of the hive is caged and either left over the top bars or removed. The candy of the virgin queen's cage is then exposed to the bees. The young queen, already half introduced, is soon liberated, and in favourable circumstances will be mated and laying within a few days. As soon as she has laid a few eggs, or even before, another virgin may be similarly introduced.

In this way a succession of fertile queens may be

---

[1] Gillet-Croix, A., *Précis d'Apiculture et Sélection des Reines*, p. 49.
[2] Perret-Maisonneuve, A., *L'Apiculture Intensive et l'Elevage des Reines*, p. 276.

obtained from a single nucleus stock, each virgin being liberated after her predecessor has begun to lay.

Maisonneuve[1] recommends the addition of 50 young bees to the hive every time a young queen is introduced in order to maintain the strength of the little stock. He speaks of the possibility, under favourable circumstances, of obtaining five or six fertile queens from a single nucleus during the course of a month. He rightly remarks, however, that the procedure demands considerable skill on the part of the queen-breeder.

3. SHAKEN QUEENLESS SWARM METHOD.

As already stated (p. 116) queenless bees confined in a ventilated box for a few hours will accept any queen dropped amongst them. This is a most reliable, if somewhat troublesome method of introducing a virgin queen of any age.

The necessary period of queenless confinement is variously given as from two to eight hours. The writer has always found six hours to be quite satisfactory.

It is usual, but not essential, to deprive the bees of food, as well as of brood and queen, during the period of imprisonment. Sladen[2] formed nuclei by shaking queenless bees into nucleus-hives provided with combs of honey, and confining them for four hours. These nuclei would then accept young virgin queens dropped into them.

It may again be emphasised that if brood is given to such nuclei a few hours after the introduction of the queen it should be sealed brood only.

[1] Perret-Maisonneuve, A., *L'Apiculture Intensive et l'Elevage des Reines*, p. 276.
[2] Sladen, F. W., *Queen Rearing in England*, 1905, p. 29.

4. HONEY METHOD.

This method has already been described in reference to the introduction of fertile queens (p. 126). Maisonneuve[1] recommends the use of liquid honey mixed with a little water. The virgin queen is dipped into this and dropped between the combs, which have been brought sufficiently close to each other to prevent her from falling to the floor of the hive. The introduction should be effected at night and without light, smoke, or noise. If the queen is more than half an hour old her reception is doubtful unless the stock has first been deprived of eggs.

This method however is troublesome and unnecessary and should not be used in preference to those now to be given.

5. WATER METHOD.

The Water Method (p. 131) is at once the simplest and most efficacious method of introducing a virgin queen of any age. It can be used in all cases where the conditions of the bees are conducive to the reception of a virgin queen. It is so quickly carried out that it is the ideal method for the busy beekeeper.

6. THE "ONE-HOUR" METHOD.

The "One-hour" Method, described on page 140, is applicable to the introduction of virgin queens. The conditions of the stock should be favourable and the introduction effected in the late evening. This method is expeditious and convenient for those who do not care to use the Water Method.

As in the case of fertile queens, virgins should not be introduced, without special precautions, to stocks which are being robbed (p. 45) or which

[1] Perret-Maisonneuve, A., *L'Apiculture Intensive et l'Elevage des Reines*, p. 272.

contain fertile workers (p. 39), and a stock should not be examined during the few days subsequent to introduction. If a queen is rejected by the bees her body will most likely be found outside the hive entrance on the following day.

CHAPTER XI

REQUEENING BY QUEEN-CELL

REFERENCE has already been made to the comparative difficulty of introducing a virgin queen. If economy of time is not important it is sometimes preferable—being easier and safer—to requeen a queenless stock by means of an almost matured queen-cell. The conditions under which this can be done need careful consideration.

In nature, except in cases of supersedure, the presence of "ripe" queen-cells is inconsistent with the presence of eggs because the old queen has left the hive with a swarm some days previously. It is therefore more difficult to persuade the bees to accept a queen-cell from another stock if eggs are present than if all the brood is sealed.

Period of Queenlessness.

If the bees have been queenless for a few days they will have begun to make queen-cells of their own and the addition of another queen-cell will make no difference to them. If the added cell be more mature than those raised by the bees, its queen will emerge first and destroy the others.

The necessary period of queenlessness before the introduction of a queen-cell is important. Writers and lecturers often speak lightly of removing the old queen and giving the bees a ripe queen-cell. Without reference to time and conditions such advice is misleading and results in failures.

We may be quite clear about this—that an unprotected queen-cell will seldom be accepted by the bees on the day when the queen is removed, except perhaps in times of nectar scarcity.[1] It may possibly be accepted on the following day, but even then it is likely to be destroyed by the bees. On the third day the prospect of acceptance is good, for by that time the bees will have made some progress with their own queen-cells. The fourth and succeeding days are even more favourable, for the youngest eggs will then be hatching and queen-cells will be well advanced.

Whilst the writer is aware that such experienced authorities as Doolittle and Sladen[2] refer to "one to two days" and "in the evening or on the next day" respectively as indicating the requisite periods of queenlessness, he has experienced so many failures in inserting queen cells on the second day that he would emphasise the desirability of waiting till the third or fourth day.

In every case the queen-cell introduced must be more mature than any that may already be in the hive.

If there is any doubt as to the age of the latter they should be destroyed before the new one is inserted. Indeed it is advisable to destroy them in any case.

Queen-cell Protectors.

Thanks to the inventiveness of Doolittle[3] we are able to introduce a queen-cell to a stock immediately after the removal of the queen. This is necessary when there are maturing queen-cells which must be allocated to nuclei or stocks without delay.

Doolittle, aware that bees destroy a queen-cell by tearing it open at its base, and that they cannot nibble through the toughened end, conceived the idea of protecting the base and sides of the cell from the bees and exposing only the end to them. He found that

---

[1] Herrod-Hempsall, W., *Beekeeping, New and Old*, Vol. I, p. 634.
[2] Sladen, F. W., *Queen Rearing in England*, 1905, p. 21.
[3] Doolittle, G. M., *Queen Rearing*, p. 54.

this prevented the destruction of the cell but did not prevent the queen from cutting away the end and emerging from it in due course. He describes his queen-cell protector in the following words:—"The cage was made by rolling a small piece of wire-cloth around a V-shaped stick, so that a small but not very flaring funnel was made, the hole in the small end being as large as an ordinary lead-pencil. After making the cage, I cut off a piece of five-eighths inch cork for a stopper, put in a nearly mature queen-cell, with the point down into the lead-pencil hole as far as it would go, when the piece of cork was put in so that the bees could not get in at the base." Later he says, "These caged cells were hung in the hives at the time the queens were removed, and in from 24 to 48 hours, according to the age of the cell, I had a nice virgin queen in the hive wherever a caged cell was hung."

Doolittle speaks of two per cent failures in the use of his cell protector. It may be remarked that he incurred some risk in inserting cells from which queens would emerge within 24 hours. If possible protected cells should be introduced at least two days before the queens are expected to emerge.

Dr. Miller[1] introduced protected ripe queen-cells to stocks without first removing the old queens. Many of the resulting young queens, he says, ultimately displaced the old ones. He adds that if the wings of the old queens are clipped their displacement can be verified. Alternatively, of course, the old queens may be marked by paint or discs.

Modifications of Doolittle's cell protector are in common use to-day. One of the best is the "Wells" or "West" Queen Cell Protector (Pl. XI, Fig. 1), which consists of a single strand of stout wire spirally wound into the form of a cone. The smaller end is open and the top closed by a small sheet of tin. An

---

[1] *L'Apiculture Française*, 1937, p. 254.

extension of the wire is made into a support by means of which the cage can be suspended between combs. A similar protector can be fashioned from perforated zinc.

An objection to some of these protector-cages, as sold, is that they are rather small. They accommodate cells built on artificial wooden cell bases quite well, but natural cells often have to be severely pared down before they can be inserted into them.

Introduction by Nursery Cage.

Queen-cells may be safely introduced in Nursery Cages. As these are closed the bees cannot injure the cells and the newly-emerged queens find an immediate provision of food which sustains them until they are liberated.

Many forms of nursery cages have been invented, the simplest being that originally made by Mr. H. Alley (Pl. XI, Fig. 2). This consists of a block of wood $2\frac{1}{2}''$ square and $1''$ thick. Through the middle of the large faces a hole of $1\frac{1}{2}''$ diameter is bored to form the queen chamber. This is covered on both sides by wire cloth. Through one of the small faces, viz., that to be at the top when the cage is in use, two holes are bored into the queen chamber, one of $\frac{3}{4}''$ diameter, with finger depressions, to accommodate a queen cell, and the other of $\frac{3}{8}''$ diameter to hold candy. Doolittle recommends a paste of granulated sugar and honey for this candy, but stiff candy with a slight admixture of pollen as described on page 110 is more suitable. The candy hole is to be covered with a small piece of tin which may be turned on a nail when it is desired to allow the bees to liberate the queen. The queen cell is suspended in the larger hole by means of the rim of its wooden cell-cup, or, if a natural one, by a cork or thin piece of wood to which it is made to adhere with a little melted wax.

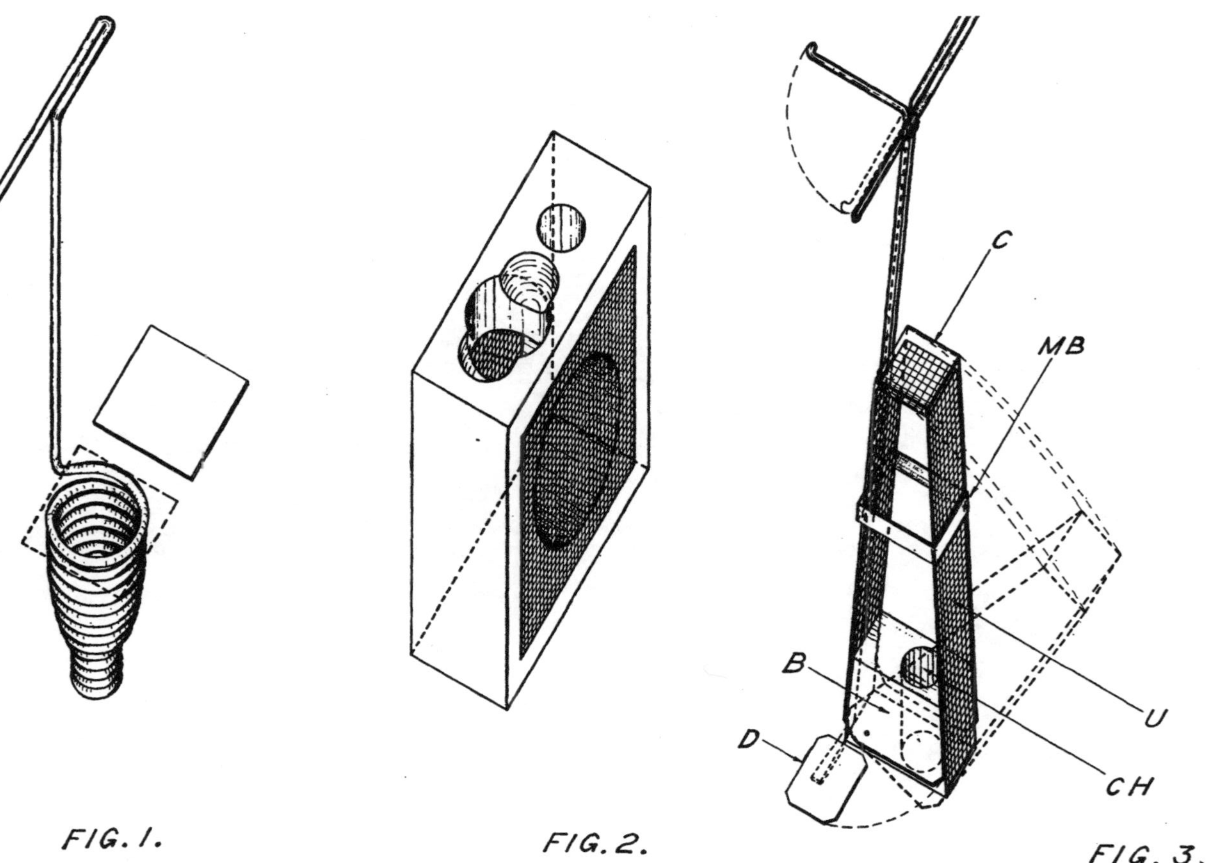
M
FIG. 1.
FIG. 2.
C
MB
B
D
U
CH
PLATE XI
FIG. 3.

A number of these cages fitted into a standard frame allows a breeder to incubate a batch of queen-cells in a strong colony. The cage however is not particularly convenient for suspension between combs when introducing by queen-cell.

A more suitable cage for introduction either of a queen-cell or virgin queen is that devised by Sladen[1] (Pl. XI, Fig. 3). This consists essentially of an inverted U-shaped piece of wire-cloth (U) tapering towards the top and fixed to a block of wood (B) at the bottom. This wire-cloth forms two sides of the cage. The other two sides are formed by an inverted U-shaped sheet of celluloid (C) riveted to the block so as to swing away from the other part of the cage when this is opened. The queen-cell is held by the wire cloth at the top of the cage, and the block, pierced with a covered candy hole (CH), provides for the food and release of the queen. When closed the parts of the cage are kept together by a metal band (MB) which slips over them. A stout and suitably bent wire soldered to the wire-cloth permits of the suspension of the cage between two combs. In some modern forms of this cage the suspending wire may be rotated so as to open a tin door (D) at the bottom of the cage in order to expose the candy to the hive bees at the right time without disturbance.

For some years the writer improvised nursery cages from 3″ lengths of a large-section bamboo rod. The ends were completed by corks—the upper one to carry the queen-cell and the lower one the candy. Small opposite segments, $\frac{3}{4}$″ in height, were cut from the middle of each cage. In the one was inserted a piece of celluloid to serve as a window and in the opposite one a piece of perforated zinc for ventilation. These cages were quite satisfactory in use. They were quickly made at negligible cost.

[1] Sladen, F. W., *Queen Rearing in England*, 1913, p. 24.

When a queen-cell is introduced by means of a nursery cage the resulting queen should not be liberated until the fourth day of queenlessness or later, and the cell should have been in the hive for at least two days.

The Gillet-Croix cage (p. 65) is specially devised for introduction by queen-cell and when used for this purpose is, in effect, a nursery cage.

Inserting Unprotected Queen-cells.

As previously stated an unprotected queen-cell may be safely inserted into a hive after two complete days of queenlessness, i.e. on the third or fourth day, or later. The cell, whether on an artificial queen-cell cup or one cut from a comb, can be placed between two combs at the feed-hole in the quilt. The space between these combs should be so adjusted that the base of the cell is slightly pressed between them so that it cannot fall to the floor, and so that the end of the cell is entirely free. The bees will quickly attach the cell to the combs.

Alternatively a wedge-shaped piece of comb containing the queen-cell and a little of the surrounding brood can be grafted into a comb of the hive to be requeened. This method is to be recommended because the queen-cell is not likely to be damaged by the knife or the bees.

Precautions to be Observed on Handling Queen-cells.

It need hardly be stated that queen-cells, whilst being removed from one hive to others, should not be exposed to the sun on a hot day, or to wind or low temperature at other times, nor should the combs on which they are built be shaken to free them of bees.

A cell built in an artificial queen-cup is easily handled. It is only necessary to hold it by the flange of the wooden cup. A natural cell has usually to be cut from the comb and this postulates some care and judgment.

If it is possible to insert a whole brood-comb containing one queen-cell into the hive to be requeened cutting is avoided and success is more probable.

To take a batch of natural queen-cells from a stock the beekeeper should provide himself with a sharp knife, a pair of scissors, and a warmed cardboard box. It is a great advantage if he has a friend to hold the combs whilst he works.

Cells at the bottoms or on the sides of the combs can be easily removed by the knife, the cut being made so as to include at least two rows of cells surrounding them. These will be helpful in suspending the cells between combs in their new hives. If the knife exposes the base wall of the queen-cell the latter will probably be destroyed when introduced. This sometimes happens when the knife is inserted between two queen cells close together. In such a case it is better to cut both out as one. They may both be introduced to a queenless stock except during the swarming season when it is prudent to damage one of them. A damaged queen-cell is useless. The bees devour the exposed royal jelly and drag the inmate out.

Some of the best cells may be built over the frame wires. These should first be carefully severed with scissors and the cells then cut out with the knife.

As each cell is removed it should be placed in a warmed box which should then be closed. For this purpose the writer uses a light cardboard box, such as contains a pound of note-paper. A medicine bottle, filled with fairly hot water, is fixed in one side of the box with neck protruding through the end. The bottle is wrapped in a layer of cotton-wool which also lines the remainder of the box. With lid closed this box keeps warm for a long time, and in it queen-cells can safely be taken for some miles in a car and kept during the time needed for distribution. The cotton-wool below and above the cells protects them from vibration and prevents them from being moved from their positions.

It is usually necessary to cut through sealed brood whilst removing queen-cells. The juices of damaged larvæ are repellent to bees, but they soon clear them away and repair the damaged cells. If this work is left to strange bees however they are sometimes tempted to destroy the queen-cells. The writer therefore suspends all the cells for an hour or two between two combs of the hive from which they were taken. The bees clean and repair the surrounding comb and do not attack their own queen-cells. These are then returned to the warm box and distributed.

Selection of Queen-Cells.

When choosing queen-cells for requeening we should reject those that are

    (1) Unusually long.
    (2) Much enlarged and thickened.
    (3) Very smooth and in the vicinity of drone brood.
    (4) Very short and crooked.

(1), (2) and (3) are unlikely to "hatch". (4) may contain ill-nourished and undersized queens.

Sladen[1] surmised that an exceedingly long cell, which is occasionally found, is "due to the larva failing to reach at the proper time the stage at which the cell should be sealed, the bees lengthening the cell".

An enlarged and thick-walled cell may contain a dead larva. A smooth-walled cell, especially in the neighbourhood of drone brood, probably contains a drone larva which will not live to emerge.

Such defective cells are seldom found in a normal stock. If the beekeeper desires to be certain of a particular cell he should select one which is not quite

[1] Sladen, F. W., *Queen Rearing in England*, 1913, p. 6.

ready for sealing and in which the larva is visible. Such a cell will be accepted more readily than a sealed one.

Other Considerations.

Requeening by means of Virgin Queens or Queen-cells should be practised only when drones are available. In England there is little chance of the queens becoming mated earlier than the beginning of May or later than the middle of September.

Nuclei used for queen-mating should have sufficient bees to maintain the normal hive temperature. A very small nucleus—except during hot weather—may not be sufficiently warm to "hatch" a queen-cell. If there is brood an inserted queen-cell should, of course, be placed in the midst of it.

Very small "baby" nuclei, such as those recommended by the American breeder "Swarthmore" and containing about half a pint of bees, are useless in the English climate. The writer's experience of them some years ago, during many patient experiments, was that in nine cases out of ten the bees absconded with the virgin queen when the latter flew for mating. Eggs, young brood, and sealed brood alike were deserted, even when only young bees were used.

The danger of desertion when queens are reared in nuclei is a real one, especially during the swarming season or very hot weather. To minimise it a nucleus stock should comprise not less than two combs containing stores and sealed brood, and should be kept in a shaded situation.

Nuclei should not be formed of bees from a hive which has developed the swarming impulse. Unless they are completely deprived of flying bees immediately before the young queens fly to mate, the latter will most likely abscond with swarms.

CHAPTER XII

SUMMARY OF THE PRINCIPAL METHODS OF QUEEN INTRODUCTION

FOR the convenience of those who desire to refer quickly to particular methods the following brief directions are given. In each case the writer's views with regard to reliability and the labour involved are indicated as follow:—

Reliable . . .	RRR.
Moderately reliable .	RR.
Less reliable . .	R.
Little labour . .	L.
Moderate labour . .	LL.
Much labour . .	LLL.

Thus the most desirable methods are marked $\dfrac{\text{RRR}}{\text{L}}$ $\dfrac{\text{R}}{}$ and the least desirable LLL. It will be observed that several of them are not to be recommended. These are included because many people use them.

A. FERTILE QUEENS

1. ORDINARY CAGING METHODS (p. 71).

$$\frac{\text{RRR}}{\text{LL}}$$

Place the cage, containing the queen with or without attendants, on its side, across the frames, and over the brood (if any), so as to expose the wire-cloth covering to the hive bees. About 48 hours later replenish the candy if necessary and expose this to the hive-bees.

A day or two later ascertain if the queen has left the cage but do not examine the combs for a week.

2. CHANTRY CAGE (p. 73). $\dfrac{RRR}{L}$

Fill both passages of the cage with stiff candy. Insert the queen, cover the cage with wire cloth, and fix a piece of stout rough paper, pierced with pin holes, over the opening leading to the longer column of candy. (If the candy is sufficiently stiff the paper is not necessary.)

Place the cage, wire cloth vertical, across the frames and over the brood, and cover with the quilts.

The queen should be liberated by the bees within three days. Examine at the end of a week.

For the use of other cages embodying the Chantry principle see pp. 74, 83, 84, 91, 92.

3. MILLER CAGE (p. 66). $\dfrac{RRR}{LL}$

Suspend the cage containing the queen only between two brood combs which need not be spaced apart. After about 48 hours expose the candy to the hive bees. The cage may be withdrawn a day later, but the combs should not be disturbed for a week.

4. PIPE COVER CAGE (p. 77). $\dfrac{RR}{LL}$

(1) Put the queen in the cage as advised on p. 79. Select a place on a brood comb where young bees are emerging and where some cells contain honey. Press the cage into the comb to the depth of the cells. Replace the comb and leave for 48 hours. Liberate the queen by raising the cage from the comb if the attitude of the

bees appears to be favourable. If it is unfavourable, cage the queen for another day.

(2) If emerging brood with honey cannot be found give the queen four or five attendants and fix the cage over sealed brood and honey.

5. Large "Press-in" Cages (pp. 80, 83).

$$\frac{RRR}{LL}$$

Cage the queen with some attendants over a patch of comb containing emerging brood, and honey. Liberate her when she has begun to lay (in four or five days) after having previously broken down all queen cells in the hive.

If the cage is placed over sealed as distinct from emerging brood, give the queen a few attendants.

6. Pieyre Type of Cage ("Pieyre", "Adam") (p. 83, 84).

$$\frac{RRR}{L}$$

Fill both tunnels with stiff candy and cover their external ends with very thin cardboard or stout paper pierced with pin holes. Fix the cage over the queen on a patch of comb containing emerging bees and some honey. The bees will liberate the queen in due course —probably after she has begun to lay. Do not disturb for a week.

If an "Adam" cage is used, first place the queen and her attendants in the smaller compartment. Fix the cage to the comb and raise the slide which allows the queen to enter the larger compartment (p. 84).

7. "WHYTE" INTRODUCING CAGE (p. 96).

$$\frac{\text{RRR}}{\text{LL}}$$

Remove from the hive a comb containing emerging brood and stores; shake it absolutely clear of bees, and put it into the "Whyte" cage. Place the cage between two combs of brood in the hive. Insert the queen through the hole in the cage lid. Replace the quilts. A week later break down queen-cells in the hive, remove the cage, take out the comb with bees and queen (the latter probably now laying), and replace it in the brood-nest.

If the hive were dequeened some days before the introduction of the "Whyte" cage, care must be taken that no queen-cells mature during the period of the new queen's confinement.

8. CYLINDRICAL CAGE (p. 85).

$$\frac{\text{RR}}{\text{LL}}$$

Place the queen in the cage without attendants. Place the cage between two combs where sealed brood and sealed honey are near together. The combs should grip the cage sufficiently to bruise a little of the honey capping so as to allow the queen to feed herself.

After 48 hours observe the attitude of the bees. If friendly, remove the lower cork of the cage and liberate the queen. If unfriendly, confine the queen for a further 24 hours.

9. RAYNOR CAGE. METHOD OF SUBSTITUTION (p. 89).

$$\frac{\text{RR}}{\text{LL}}$$

Close the lower door of the cage and suspend it between two brood combs. Open the top door and insert the old queen. Close the cage and leave it exposed to the bees for 12 hours. Remove the cage, open the top door and smoke the cage gently until the old queen walks out. Place the new queen in the cage at once. Close the cage and expose it between the combs for a further 24 hours. Open the lower door without removing the cage, and so liberate the queen without disturbance, preferably after dark.

10. PRESENTING AT ENTRANCE (p. 115).

$$\frac{R}{L}$$

Dequeen the hive. About 6 hours later present the cage containing the new queen (without attendants) to the distressed bees at the hive entrance. If the bees appear to welcome her, liberate her and let her run into the hive.

11. SHAKEN QUEENLESS SWARM METHOD (p. 116).

$$\frac{RRR}{LLL}$$

Shake queenless bees into a ventilated box. Keep in a shaded place for about 6 hours. Drop the new queen into the box and leave for a further hour or two. The clustered artificial swarm can then be shaken into a hive provided with stores but no unsealed brood.

12. RAYNOR'S SHAKEN BEES METHOD (p. 117).

$$\frac{RRR}{LLL}$$

Remove the stock temporarily to a new position and place an empty skep on a floor in the original hive position. Shake all bees from the combs in front of the skep and remove the old queen as they run into it.

Replace the hive in its original position. Shake the bees from the skep on to a cloth in front of the hive. As the terrified bees run in drop the new queen amongst them.

13. PARTLY SHAKEN BEES METHOD (p. 118).

$$\frac{R}{LL}$$

Dequeen the hive. Shake three combs of bees taken from the middle of the hive on to a cloth placed in front of the hive. As the bees run in drop the new queen amongst them.

14. NUCLEUS METHOD (p. 118).

$$\frac{RR}{LL}$$

The hive should be queenless for at least 24 hours. If longer, queen-cells should be destroyed.

At dusk remove quilts from both hive and nucleus and expose the bees to the air for 5 minutes. Insert the nucleus at the back of the hive so that the comb face carrying the queen is next to the hive wall. Cover the hive without disturbance.

15. SMOKE METHOD (p. 122).

$$\frac{R}{LL}$$

Use in the evening when all the bees are at home.

Reduce the hive entrance until only sufficient to admit the nozzle of the smoker. Puff smoke into the hive, gently at first, then more vigorously. When the bees roar with fright let the queen run in and close the entrance for ten minutes. Re-open and enlarge the small entrance. After an hour restore the full entrance.

16. Uniting by Paper Method (p. 123).

$$\frac{RRR}{LL}$$

A queen-right nucleus is to be united to a queenless stock. Strengthen the nucleus with sealed brood some days before uniting—especially if it is rather weak. Destroy queen-cells, if any, in the queenless stock and unite as described on p. 124.

17. Simmins "Fasting" Method (p. 124).

$$\frac{RRR}{L}$$

The hive should be queenless for about 6 hours (or more—v. p. 125), and the introduction effected after darkness has set in.

Place the queen in a match-box and keep this in the pocket for 30 minutes. With the least possible disturbance raise the hive quilt and give a preliminary puff of smoke. Place the match-box, opening downwards, over the space between two frames, push it open with the finger, and quickly replace the quilts.

The introduction can be effected with less disturbance viâ the feed-hole.

18. Immersion in Honey Method (p. 126).

$$\frac{RR}{L}$$

Warm a little honey slightly diluted with water. Immerse the queen in it and drop her between two combs which are near together. (If the queen falls to the floor she may die before the bees clean her.)

19. FLOUR METHOD (p. 127).

$$\frac{R}{L}$$

In favourable circumstances, place the queen in flour so that she is thoroughly covered with it and put her amongst the bees (e.g. through the feed-hole).

20. CHLOROFORM METHOD (p. 127).

$$\frac{RRR}{L}$$

Towards dusk take two pieces of corrugated paper and on each pour about two-thirds of a tea-spoonful of chloroform. Insert one piece into the hive at the entrance and the other under the quilts at the back of the hive. Close the entrance with a duster for three minutes. Remove the pieces of paper, drop the queen amongst the bees, replace the quilts quickly, and re-open wide the hive entrance.

21. "SNELGROVE" WATER METHOD (p. 130).

$$\frac{RRR}{L}$$

(*a*) Dequeen the stock. Take the new queen by the wings, dip her into luke-warm water, moving her to and fro two or three times to make sure she is thoroughly wet, and at once let her run amongst the bees on the combs. Cover the hive without disturbance and do not examine for several days.

(*b*) Dequeen the hive. Place the new queen in a match-box. Immerse the box in tepid water opening it a little until the water nearly fills it. Agitate it gently for about five seconds and then place it, opening downwards, over a seam of bees. Leave the match-box in position and cover it with the quilts or otherwise.

The introduction in both (*a*) and (*b*) can conveniently be effected at the feed-hole in the quilt, and should be carried out during the evening if possible.

22. "SNELGROVE" SWARM-CONTROL BOARD METHOD (p. 132).

$$\frac{RRR}{LL}$$

Shake bees from combs of the ripest brood. Place these combs in a brood box and put this over an excluder placed on the original stock. The next day substitute the swarm-control board for the excluder, leaving one upper exit in the board. After one day of activity the old bees will have left the upper box, and to the young bees remaining a new queen can be introduced by any good method with absolute safety.

23. SNELGROVE "ONE-HOUR" METHOD (pp. 140, 141).

$$\frac{RRR}{L}$$

Towards dusk dequeen the hive and place from 20 to 30 of the bees in a safety match-box. Keep it in the vest pocket for about 10 minutes. Put the new queen into the box with the bees. Return it to the pocket for 50 minutes (or longer up to a maximum of $1\frac{1}{2}$ hours). After a preliminary puff of smoke, place the box, opening downwards, under the quilt or over the feed-hole, and push it open with the finger to allow the queen to walk down between the combs. Cover quickly and without noise and leave the hive undisturbed for a few days.

24. INTRODUCTION TO A STOCK CONTAINING A DRONE BREEDING QUEEN (pp. 50 and 139).

$$\frac{RRR}{L}$$

Remove the queen and give the stock a comb of brood in all stages. A day later, in the evening, introduce the new queen by the "One-hour" method (p. 140).

25. INTRODUCTION BY CAGING TO A STOCK CONTAINING FERTILE WORKERS (p. 99).

$$\frac{RR}{LL}$$

Insert a comb of unsealed brood into the stock two days beforehand. Give the new queen a few attendants from the fertile worker stock (p. 135) and place the cage over the frames in the usual way. On the third day, towards night, break down all pseudo-queen cells and expose the candy of the cage to the hive-bees. The queen will be liberated during the night. Do not disturb the stock for a few days.

26. DIRECT INTRODUCTION TO A STOCK CONTAINING FERTILE WORKERS (pp. 100, 140).

$$\frac{RRR}{L}$$

Insert a comb of brood—some unsealed, into the stock one or two days previously. Break down pseudo-queen cells and towards night introduce by the One-hour method (p. 140). Leave undisturbed for several days.

N

27. Introduction to Bees Long Queenless (p. 101).

$$\frac{RRR}{L}$$

(*a*) *In early spring.* Introduce by the "One Hour" method. Brood may be inserted previously but this is not usually available or necessary.

(*b*) *During summer and autumn.* Potentially fertile workers may be present—therefore introduce some young brood a day or two before introducing the queen by the Water method (p. 130), or by the One-hour method (p. 140).

28. To Introduce a Valuable Queen without Risk.

(*a*) Use a Whyte Introducing Cage (p. 96), or

(*b*) Remove the stock to a new stand. Remove a comb with the old queen and place it in another hive on the old stand. Leave for 24 hours, or longer if weather is bad, to allow the old bees to leave the original stock.

Introduce the new queen to the young bees late in the evening by any one of the most reliable methods here given.

B. VIRGIN QUEENS

29. Caging Method (p. 168).

$$\frac{RR}{LL}$$

Sealed brood only, with stores, should be present. Keep the stock queenless for at least one day. Provide the cage with candy containing a little pollen. If a "press-in" cage is used fix it over sealed brood, honey and pollen. After 3 days break down queen-cells, expose the candy of the cage to the hive-bees, or, in the case of a "press-in" cage, liberate the queen.

30. SHAKEN QUEENLESS SWARM METHOD (p. 170).

$$\frac{RRR}{LLL}$$

Shake the bees into a ventilated box. Keep them confined in a dark place for 6 hours. Drop the queen amongst them. An hour or two later, or when the bees have clustered as a swarm, shake them on to combs containing stores. On the following day give them a little sealed brood, but no eggs or young brood.

If preferred, the ventilated box may contain combs of stores but no brood.

31. HONEY METHOD (p. 171).

$$\frac{RR}{L}$$

See No. 18. Sealed brood only, with stores, should be present. The stock should be queenless for at least three days. Break down queen-cells before introducing the queen.

32. WATER METHOD (p. 171).

$$\frac{RRR}{L}$$

Sealed brood only, if any, with stores, should be present. The stock should be queenless for at least three days. Before introducing the queen (p. 130) break down all queen-cells.

33. One-Hour Method (p. 171).

$$\frac{RRR}{L}$$

Sealed brood, if any, with stores, should be present. Before introducing the queen break down all queen-cells (p. 140).

34. Introduction of Unprotected Ripe Queen-Cell (p. 180).

$$\frac{RRR}{L}$$

The stock should be queenless for two whole days or more and should preferably have some brood. If queen-cells in the stock are advanced break them down before introducing the selected queen-cell.

35. Introduction of Protected Queen-Cell (p. 174).

$$\frac{RRR}{L}$$

Remove the old queen and at the same time, or soon afterwards, place a sealed but immature queen-cell in a cell-protector. Suspend this between the combs. The cell should be in the hive for at least two days before the queen is due to emerge.

CONCLUSION

THE reader of this book will realise that to be successful in Queen Introduction it is not sufficient merely to be familiar with correct procedure in the actual introduction of queens. It is equally important to study, and to modify when necessary, the conditions of the stocks which are to receive the queens, and to exercise judgment as to the most suitable method to be employed in every case.

The commonest causes of failure are:—

(1) The presence of unsuspected queen-cells or virgin queens.
(2) The presence of fertile workers, whether laying or not.
(3) Unsuitable conditions of the brood.
(4) Robbing.
(5) The premature release of caged queens.
(6) The introduction of queens which have travelled for many days, without proper precautions.
(7) The release of queens during the day-time, i.e., when bees are flying.
(8) Premature examination after introduction.

Of the many precautions to be observed the writer would emphasise the following, at the risk of repetition, as specially conducive, although not always essential, to success:—

(1) The queenless stock should have brood of the appropriate age.

(2) The queen should be liberated in the hive at
    night.
(3) A specially valuable queen should be introduced
    to young bees.

If failures occur the reasons for them should be
found in Chapters III and IV.

It is the sincere wish of the writer that this book
will encourage many to undertake their own requeening
and that the advice and directions he has given will
enable them to do so with confidence and success.

BIBLIOGRAPHY

ADAM, REV. BRO.	*Article on Queen Introduction, Somerset Beekeepers' Annual Report*	Bridgwater	1934
ALPHANDÉRY, E.	*Le Livre de l'Abeille*	Paris	1937
ALPHANDÉRY, E.	*Traité Complet d'Apiculture*	Paris	1931
BERTRAND, E.	*La Conduite du Rucher*	Paris	1937
BETTS, A. D.	*Practical Bee Anatomy*	Benson	1923
BETTS, A. D.	*The Bee World*	Foxton, Cambridge	1938
CHESHIRE, F. R.	*Bees and Beekeeping, Vol. II (Practical)*	London	1886
COTTON, W. C.	*My Bee Book*	London	1842
COWAN, T. W.	*The British Beekeeper's Guide Book*	London	1913
DOOLITTLE, G. M.	*Queen Rearing*	New York	1899
FREUDENSTEIN, DR. K.	*Wege zur Völker Vermehrung. Deutscher Imkerführer*	Berlin, Juli	1937
GILLET-CROIX, A.	*Précis d'Apiculture et Sélection des Reines*	Bertrix	1924
GILMAN, A.	*Practical Bee-Breeding*	London	1930
HALLEUX, D.	*Le Livre de l'Apiculteur Belge*	Huy	1911
HERROD-HEMPSALL, W.	*Beekeeping, New and Old, Vol. I.*	London	1930
HOMMEL, R.	*Apiculture*	Paris	1927
HUBER, F.	*Nouvelles Observations sur les Abeilles. 2me Edition*	Paris and Geneva	1814
HUISH, R.	*A Treatise on the Nature, Economy, and Practical Management of Bees*	London	1817

HUNTER, J.	*A Manual of Beekeeping*	London	1884
HUTCHINSON, W. Z.	*Advanced Bee Culture*	Flint, Michigan	1905
ILLINGWORTH, L.	*A Safe and Simple Method of Requeening a Bad-tempered Stock* *Bee World*, January 1940	Foxton, Cambridge	1940
KEEN, G. H.	*Introducing Queen Bees*	West Maitland, N.S.W.	1938
KOCH, K. K.	*Was treibt zum Schwarmen?* *Deutscher Imkertag*	Görlitz	1932
LANCINI, V.	*Questiti e Risposte* *L'Apicoltore d'Italia*	Ancona	1938
LANGSTROTH, L. L.	*The Hive and Honey Bee*	Hamilton, Illinois	1911
LEUENBERGER, F.	*Les Abeilles. Anatomie et Physiologie (Traduction Jaubert)*	Paris	1929
MANLEY, R. O. B.	*Honey Production in the British Isles*	London	1936
MILLER, DR. C. C.	*Fifty Years Among the Bees*	Medina, Ohio	1911
PELLET, F. C.	*Productive Bee-keeping*	Philadelphia,	1918
PELLET, F. C.	*History of American Bee-keeping*	Ames, Iowa	1938
PELLET, F. C.	*Practical Queen Rearing*	Hamilton, Illinois	1917
PERRET-MAISONNEUVE, A.	*L'Apiculture Intensive et l'Elevage des Reines*	Paris	1926
PURCHAS, S.	*A Theatre of Political Flying-Insects*	London	1657

RAYNOR, G.   *Queen Introduction. The Ligurian Bee.* A paper read to the B.B.K. Association. January, 1880   London   1884

REAUMUR, K. A. F.   *Mémoires pour servir à l'Histoire des Insectes.* Vol. V   Paris   1740

ROOT, A. I. & E. R.   *ABC and XYZ of Bee Culture.*   Medina, Ohio   1929, 1940

SELLNER, KARL   *Natürliche Königinnenzucht* *Bienen Vater*, Juni, 1937   Vienna   1937

SIMMINS, S.   *A Modern Bee Farm*   London   1940

Sladen, F. W.   *Queen Rearing in England*   London   1905, 1913

SMITH, JAY   *Queen Rearing Simplified*   Medina, Ohio   1923

SMITH, JAY   *The New Jay Smith Cage Bees and Honey*   Alhambra, California   1938

SNELGROVE, L. E.   *Swarming, Its Control and Prevention*   Bleadon   1935

SNODGRASS, R. E.   *Anatomy and Physiology of the Honey Bee*   New York   1925

TARR, DR. H. L. A.   *Brood Diseases in England. Rothamsted Conferences*, XXII   Harpenden   1936

THE TIMES BEEMASTER   *Beekeeping*   London   1864

THORLEY, J.   *The Female Monarchy*   London   1744

TUENIN, T. A.   *Concerning Laying Workers* *Bee World*, Nov.—Dec. 1926   Foxton, Cambridge   1926

TURNER, N. H.   *The Care and Management of the Queen*
Scottish Bee Journal,
Oct. 1938   Arbroath   1938

VON RHEIN, W.   *Über die Entstehung des weiblichen Dimorphismus im Bienenstaate*
Arch. f. Entwicklungsmechanik, 129. 4.   Berlin   1933

WEDMORE, E. B.   *A Manual of Beekeeping*   London   1932

PERIODICALS

The Bee World	Foxton, Cambridge
The British Bee Journal	London
Bee Craft	Sidcup
The Scottish Beekeeper	Arbroath
Bienen Vater	Vienna
Deutscher Imkerführer	Berlin
The American Bee Journal	Hamilton, Illinois
Gleanings in Bee Culture	Medina, Ohio
La Gazette Apicole	Montfavet, France
L'Apiculture Française	Sainte-Soline, France
Bees and Honey	Alhambra, California
Australasian Bee Journal	Maitland, N.S.W.
L'Apicoltore D'Italia	Ancona, Italy

INDEX